New Directions in
Sorption Technology

New Directions in Sorption Technology

Edited by

George E. Keller II
Union Carbide Corporation
South Charleston, West Virginia

Ralph T. Yang
State University of New York at Buffalo
Buffalo, New York

Butterworths
Boston   London   Singapore   Sydney   Toronto   Wellington

Library of Congress Cataloging-in-Publication Data
New directions in sorption technology / edited by George E.
  Keller II, Ralph T. Yang.
       p.   cm.
  Selected papers from the Symposium on Improved
Adsorbents and Adsorption Processes and an Award
Symposium on Separation Science and Technology, held at
the 193rd National Meeting of the American Chemical Society
in Denver, 1987.
  Bibliography: p.
  Includes index.
  ISBN 0-409-90183-0
  1. Adsorption—Congresses.   I. Keller, George E.   II.
Yang, Ralph T.   III. Symposium on Improved Adsorbents
and Adsorption Processes (1987 : Denver, Colo.)   IV. Award
Symposium on Separation Science and Technology (1987 :
Denver, Colo.)   V. American Chemical Society. Meeting
(193rd : 1987 : Denver, Colo.)
TP156.A35N48   1989
660.2 '8423—dc19   89-30927

British Library Cataloguing in Publication Data
New directions in sorption technology.
  1. Chemical engineering. Adsorption
  I. Keller, George E.   II. Yang, Ralph T.
  660.2 '8423

  ISBN 0-409-90183-0

Butterworth Publishers
80 Montvale Avenue
Stoneham, MA 02180

10 9 8 7 6 5 4 3 2 1

Printed in the United States of America

CONTENTS

CONTRIBUTORS

Heinrich Amlinger
Leybold-Heraeus GMBH
D-6450 Hanau 1
West Germany

P.E. Barker
Department of Chemical
  Engineering and Applied
  Chemistry
Aston University
Birmingham B4 7ET
United Kingdom

Charles H. Byers
Chemical Technology Division
Oak Ridge National
  Laboratory*
Oak Ridge, Tennessee

Giorgio Carta
Department of Chemical
  Engineering
University of Virginia
Charlottesville, Virginia

Steven M. Cramer
Bioseparations Research Center
Department of Chemical
  Engineering
Rensselaer Polytechnic Institute
Troy, New York

G. Ganetsos
Department of Chemical
  Engineering and Applied
  Chemistry
Aston University
Birmingham B4 7ET
United Kingdom

Jeffrey H. Harwell
Institute of Applied
  Surfactant Research and
School of Chemical Engineering
  and Materials Science
University of Oklahoma
Norman, Oklahoma

J. Hearn
Department of Physical
  Sciences
Trent Polytechnic
Nottingham
United Kingdom

Friedrich G. Helfferich
Department of Chemical
  Engineering
Pennsylvania State University
University Park, Pennsylvania

W. Hinz
Leybold AG
Oerlikoner Strasse 88
CH-8057 Zurich
Switzerland

Somesh C. Nigam
University of Michigan
Department of Chemical
  Engineering
Ann Arbor, Michigan

Z.J. Pan
Department of Chemical
  Engineering
State University of New York at
  Buffalo
Buffalo, New York

*Operated by Martin Marietta Energy Systems, Inc., for the U.S. Department of Energy, under contract DE-AC05-840R21400.

J.A. Ritter
Department of Chemical
  Engineering
State University of New York at
  Buffalo
Buffalo, New York

Bruce L. Roberts
Surfactant Associates Inc. and
Institute for Applied Surfactant
  Research
University of Oklahoma
Norman, Oklahoma

John F. Scamehorn
Surfactant Associates Inc. and
Institute for Applied Surfactant
  Research
University of Oklahoma
Norman, Oklahoma

H. Schilling
Leybold AG
Wilhelm-Rohn-Strasse 25
D-6450 Hanau 1
West Germany

P.L. Smelt
Department of Physical Sciences
Trent Polytechnic
Nottingham
United Kingdom

Jin-Eon Sohn
Institute of Industrial Science
University of Tokyo
7-22-1 Roppongi Minato-ku
Tokyo
Japan

Guhan Subramanian
Bioseparations Research Center
Department of Chemical
  Engineering
Rensselaer Polytechnic Institute
Troy, New York

Motoyuki Suzuki
Institute of Industrial Science
University of Tokyo
7-22-1 Roppongi Minato-ku
Tokyo
Japan

Henry Y. Wang
University of Michigan
Department of Chemical
  Engineering
Ann Arbor, Michigan

M.C. Wilkinson
CDE Porton Down
Salisbury
United Kingdom

PREFACE

Adsorption (sometimes called sorption to include dissolution of sorbates into certain kinds of sorbents) has become increasingly important in both the traditional chemical and petrochemical industries as well as in developing industries such as biotechnology. Sorption processes range in size from those used in analytical chemistry to massive units processing millions of cubic feet of gas or liquid per day. In acknowledgment of the importance of sorption, a full-day Symposium on Improved Adsorbents and Adsorption Processes was held at the American Chemical Society's 193rd National Meeting in 1987 in Denver. At the same meeting, an Award Symposium on Separation Science and Technology was held to honor Professor F.G. Helfferich for his contributions to the areas of chromatography (a type of sorption process) and ion exchange.

This book includes selected papers presented at the two symposia as well as added papers. All papers were subjected to critical peer review, and the revisions are published here. Our objective is to report current developments in sorption technology from which new directions will emerge. The coverage is selective rather than exhaustive. Four areas are examined in this book: new sorbents, chromatography, pressure swing adsorption, and bioseparations involving sorption. Keller provides an overview concentrating on the technical maturity of various sorption processes.

Helfferich reviews the powerful concept of coherence, which he developed in 1963 to describe the behavior of concentration waves of a multicomponent mixture in a chromatographic column. He suggests broader applications of this concept to the dynamics and optimization of industrial chemical processes. His ideas are echoed by Harwell's treatise on the complex system of chromatographic movement of mixed surfactants. Here the predictive power of the coherence concept is demonstrated, including unexpected behaviors at the critical micelle concentration.

The fundamental sorption properties and the characterization of two new sorbents are described by two groups: Suzuki and Sohn on activated carbon fibers manufactured by a Japanese company and Hearn et al. on macroreticular resins manufactured by one British and three U.S. companies. The unique advantages of these new sorbents for certain applications are shown by experimental results.

In the area of continuous steady-state chromatography, Carta and Byers describe their successful operation of the annular rotating bed chromatograph, which is the only remaining hope for the moving-bed chromatograph. Results of the separations of sugars, metal ions, and amino acids are given. An authoritative review of liquid chromatographic systems, both batch and continuous, is given by Barker and Ganetsos. The chapter includes many important bioseparations. Chapters on separation and purification of biomolecules naturally follow those on chromatography, which is the workhorse for bioseparations. A useful account of liquid chromatographic systems is given by Cramer and Subramanian. The emphasis of their chapter is, however, on displacement chromatography, on which they have had extensive experience. Nigam and Wang describe their development of a new type of sorbent for bioseparations: an immobilized affinity sorbent. The sorbent is encapsulated in a membrane that provides an additional kinetic selectivity for separation.

Three chapters related to pressure-swing adsorption (PSA) for gas separations are included. Pan et al. demonstrate the underlying principles for kinetic separation by PSA through their experimental results on nitrogen production from air using 4A zeolite. Amlinger discusses the sizing of vacuum pumps for desorption in PSA and shows how pumping systems can be optimized for minimizing energy consumption. Landfill gas (methane-carbon dioxide) separation is accomplished by kinetic PSA separation using molecular sieve carbon. A new adsorption process for purifying landfill gas before its entrance to the PSA system at a commercial plant in West Germany is described by Schilling and Hinz. Finally, a promising technique for regenerating spent activated carbon in aqueous applications is suggested by Roberts et al. In this technique, called surfactant-enhanced carbon regeneration, a concentrated surfactant solution is passed through the spent bed, and the adsorbate desorbs and is solubilized into micelles, which are highly concentrated in desorbed adsorbate.

We wish to express our appreciation to the authors and reviewers and to Butterworth Publishers for their patience and fine work that made this book a reality.

George E. Keller II
Ralph T. Yang

Chapter 1

Coherence: Power and Challenge of a New Concept

Friedrich G. Helfferich

An overview of the coherence concept is given, outlining its nature, origin, predictive power, past and potential future applications, and challenges within the broader context of evolution of dynamic theory.

This presentation does not describe new hard-core developments or results, nor is it a comprehensive review. It attempts to provide a broad-brush overview of the concept of coherence, of what I meant when I chose that word twenty-five years ago, of what it has done for us and what I believe it still can do to bring us beyond where we stand today.

What Is Coherence?

For proper perspective, a good start may be to outline what coherence is and is not. It is certainly not one of the great ideas that have changed the picture we draw of our world. To name only some of the most prominent: the law of gravity, the first and second laws of thermodynamics, the quantum, the concept of relativity. Coherence is much more mundane in that it changes not one iota in the mathematical description of our world. Anyone looking for such exalted content tries to find what is not there.

Coherence is also not a method, technique, or device invented to make difficult mathematical problems more tractable. In this class would be, for example, the Laplace transform, matrix algebra, and the method of characteristics.

What coherence does is identify a particular facet, a fact of life, that had all the time been inplicit in our equations, to formulate it explicitly, and to forge it into a concept that helps us to understand phenomena, to recognize cause and effect, to predict without much need for calculation - something that makes us <u>know</u> what will happen. To name examples from this class of ideas: equilibrium, steady state, and entropy. We could work with our differential equations of motion, heat transfer, reaction, etc., and obtain the same results without ever having defined equilibrium or steady state. We could do thermodynamics without ever defining entropy. Yet these concepts facilitate our calculations and greatly help us in understanding what goes on.

The closest cousins to coherence are, indeed, equilibrium and steady state. Both are states to which a system settles down if not further disturbed (with exceptions in both cases). So is coherence. Equilibrium is restricted to a closed system. For the steady state, we allow the system to be open, but restrict it to fixed boundaries and constant boundary conditions. Coherence is not so restricted. In fact, it can be defined so that it includes equilibrium and steady state as special cases, but it also covers deviations from those states - typically, the propagation of perturbations through systems otherwise at equilibrium or steady state.

The idea that non-equilibrium, non-steady state systems shake themselves down to simple "modes" that are independent of the initial conditions is not new at all. In mechanics we have the pendulum, the harmonic oscillator, the musical string, with modes mathematically characterized by eigenvalues. All around us we see many such manifestations, from ocean swell to a flag in high wind: The world we live in is full of eigenvalues. It is just in physical chemistry and chemical engineering that our preoccupation with equilibrium and steady state has never before let us specifically and explicitly identify such dynamic behavior. Coherence does just that.

A Simple Example

To see how coherence may manifest itself in the world of physical chemistry and chemical engineering, consider a very simple physical system (1). Imagine an empty tube into which a water-oil mixture is introduced at the bottom, with no turbulence in the tube and with an oil lighter than water (see Figure 1). As oil in the tube will rise faster than water, a layer of oil will form and grow at the top of the rising column of liquid. The single change of inlet conditions (the start of oil-water flow) is propagated not as a single wave, but as two waves with different velocities - "waves" being defined as traveling variations of composition. In the distance-time plane, with distance from inlet and time as coordinates, the waves trace different trajectories (see Figure 2). If the entering fluid had contained additional immiscible phases, there would have been more than two waves. The generation of a whole spectrum of waves by a single perturbation and the formation of new zones of compositions that could not have been obtained by mixing the initial and injected fluids, such as the oil zone in our tube, are typical of multicomponent coupled dynamic systems. In the language of coherence theory, the original perturbation - the wave between air in the tube and the oil-water mixture just beginning to enter - is "noncoherent" and is resolved into two "coherent" waves: between air and oil, and between oil and oil-water. The original, noncoherent wave cannot travel as such because the oil layer immediately starts to form. In contrast, the two coherent waves travel maintaining their identities and will keep doing so until exiting the tube or being disturbed by some new manipulation.

A picture useful in more complex situations is to view a noncoherent wave as a superposition (not an additive one, though) of coherent waves which will separate from one another as they travel.

This very simple example also yields immediately a necessary and sufficient condition for a wave to be coherent. Any multicomponent wave can be viewed as a composite of single-component waves. For instance, the wave between oil and the oil-water mixture entails composition variations of both oil and water and so can be regarded as a composite of an oil wave and a water wave. For the composite wave to be coherent, that is, to travel without splitting up, the wave velocities of water and oil must obviously be the same. Equating these wave velocities leads to an eigenvalue problem, with eigenvalues characteristic of the velocities of coherent waves, and with eigenvectors characteristic of the composition variations across such waves.

The example can also serve to show one essential point that can be used to define coherence. Since the wave velocities with respect to all components must be equal at any point in a coherent wave, a complete set of values of dependent variables coexisting at such a point in space and time will travel jointly, in the same direction and at the same speed, and so remain in each other's company. A possible definition of coherence is the conservation of such sets. This definition includes equilibrium and steady state, conditions in which no value moves with time, so no existing set is ever broken up.

In this simple example, the coherence condition may appear trivial, yet it provides the key to the mathematical treatment and leads directly to the predictive tools coherence theory employs.

Tools

The coherence condition constrains the composition variations across coherent waves to eigenvectors, to definite directions in the composition space (that is, the space with the dependent variables as the coordinates, usually species concentrations, fractional phase volumes, or the like, also called "hodograph space" in the more rarified language of higher mathematics). In this space, curves meeting the coherence condition can be mapped as a grid. Several examples of such path grids in three- and four-component systems, taken from practical applications of the concept to ion exchange and enhanced oil recovery, are shown in Figures 3 to 7. In each case, the space is completely covered by an infinite number of curves, of which only some at regular intervals are shown. Any composition variation along a curve of the grid is coherent, any other is not. Coherence theory tells us that a system, in response to a perturbation, will shake itself down to a coherent state, as happened in our tube - that is, it will settle down to composition variations that follow the curves of the grid, in a sequence given by some very simple selection rules that are obvious in practice.

The important point here is that the grid is given by the type of basic equations and parameters of the system - the separation factors in adsorption and ion exchange, the relative volatilities in vapor-liquid systems, the relative permeabilities in multiphase flow in porous media, etc. - and is independent of the compositions of the initial and injected fluids. Therefore, for given components, the grid can be established once and for all and then be used to predict responses to any arbitrary initial and injection conditions. One may look at the grid as constituting the "groves" into which a system wants to settle, or as a road map on which a system's response can be mapped like a vacation trip. The combination of the distance-time diagram with the route which a system traces in the composition space provides a complete and compact quantitative description in space and time.

Calculation of an exact grid requires solutions of the eigenvalue problem at closely spaced points in the entire composition space. This is, of course, laborious. The art in applying coherence theory in practice is to realize how key features of the basic equations of equilibrium and motion are reflected in the topology and other properties of the grid. A single example may suffice to illustrate such a generation of an approximate path grid (5). In the grid of the brine-oil-surfactant system in Figure 6, paths exist only in the two-phase region (any composition variation in the single-phase region being automatically coherent). For the type of phase behavior as in the figure and with normal behavior of relative permeabilities, the path topology is always as shown. The straight paths are the tie lines of the two-phase region; the path originating from the plait point is the "equivelocity curve," easily calculated as the locus of equal velocities of the two phases; the points at which a curved path is tangential to a straight one are singularities, easily located as satisfying the condition that the two eigenvalues be equal. With all these features taken from the phase-equilibrium information or accurately calculated with simple algebra, an approximate grid is readily filled in without need to solve the eigenvalue problem at any point.

Once this art of grid generation is mastered, the practitioner can quickly predict at least qualitatively the response of his system to any perturbation, with little or no further recourse to mathematics. In fact, he can outguess a computer by predicting fairly accurately and with just pencil and paper the results of numerical simulations in less time than normally required for turn-around on most machines.

When to Resort to Coherence?

For systems as simple as the example shown in Figure 1 there is neither need nor good reason to invoke coherence. The results can be obtained as easily or more so with a conventional approach.

All that coherence can contribute here is a proof of uniqueness of the solution, as a purist may demand. Not so if subsequent perturbations generate new sets of waves, and waves of different sets begin to interfere with one another. The most complex cases are those involving gradual starting perturbations (as in gradient elution in chromatography) or interference of diffuse coherent waves with one another, producing finite regions of noncoherence in the distance-time plane (see Figure 8). No pre-coherence approach has been capable of handling such cases, except by blind numerical step-by-step integration in space and time of the differential equations or with other essentially equivalent methods. In contrast, coherence theory provides immediate qualitative answers as well as simple quantitative approximations (3). In essence, when two waves of different eigenvalues meet, local noncoherence arises temporarily and is resolved into new coherent waves in the same way as is a noncoherent perturbation at the inlet: A noncoherence does not know or care whether it was generated at start and at the inlet or arose later and within the system from wave interference. Also, except for differences in sharpness of the resulting coherent waves, it does not matter whether the noncoherent wave being resolved is sharp or diffuse. The general picture mentioned earlier, that of a noncoherent wave as a superposition of coherent waves that will separate upon propagation, still is valid. In fact, the mathematical proof for attainment of coherence from arbitrary starting conditions (6) is more easily given for a gradual then a discontinuous starting variation.

A Critical View

This overview would not be complete without addressing the concerns of critics and skeptics. There is the applied mathematician, to whom the development of coherent patterns seems trivial and not worth making a great fuss about - but he seems unaware of the predictive power of the concept in practice. Then there is the practical engineer, who well realizes the implications but has that nagging doubt whether all this can be true - if it were, why has it not long been in our textbooks? More specifically and factually, a chemist or engineer hearing about coherence for the first time is apt to ask: "In the real world, do systems actually become coherent, and if so, how long does that take?"

   To those two questions there are direct answers. First, in our ideal oil-water tube, granted all kinds of simplifying premises, behavior is coherent from the start. The same is true for comparable systems involving adsorption, ion exchange, multiphase flow, etc. But resolution into coherent waves is no longer instantaneous if the perturbation at the inlet is gradual. How long a time attainment of coherence then requires depends on several factors, the most important one being the differences in eigenvelocities of the coherent waves. In a grid region where the curves of different families are almost parallel, the velocity differences are small and resolution will take a long time. At a point where the curves become tangential, it will take forever. Worse, once the simplifying premises such as ideal flow behavior or infinite inter-phase mass transfer rates are relaxed, coherence is no longer attained at all, it is only approached asymptotically. So, in our real world, strictly speaking, nothing is ever coherent. Of course, this is disappointing. It might even tempt us to relegate coherence to the role of an academic toy rather than a practical tool.

Allow me to approach the answer indirectly and on a personal note. I well remember when I was five years old and my mother explained equilibrium to me. In the first place, I did not believe what she said was true. And if so, it seemed a fascinating idea, but one without much practical value because the world I saw was push and pull, with hardly anything at "equilibrium." And even if something ever was, nothing happened any more and so the whole thing had become kind of pointless. About eight or ten years later, my first reaction to the concept of steady states was similar. No wonder that coherence at first encounter tends to evoke in almost anyone just such an intuitive response.

Of course, such critique notwithstanding (rather well taken for a five-year old, I still feel), we use equilibrium and steady state extensively. Indeed, physical chemistry and chemical engineering without them would be unthinkable. It does not matter that in the real world, strictly speaking, nothing is ever in equilibrium or steady state as both are only approached asymptotically. In fact, we use equilibrium even where we know fully well that our system is and will remain far from it. Any engineer will appreciate just one example, that of a countercurrent gas absorption column. In principle, there can be no equilibrium between the bulk phases because equilibrium curve and operating curve cannot coincide - and if local equilibrium were forced into the equations, the immediate but hardly useful answer would be that an infinitesimally short column should then suffice to do any specified job.

What equilibrium does for us in all these situations is two things: It provides key parameters in our mathematical equations, and it tells us conceptually the direction in which the system wants to move - toward equilibrium - and so lets us better understand the physical forces at work, identify cause and effect, become proficient at predicting qualitatively what will happen under given circumstances, acquire judgment, conceive better solutions. The practical importance of this cannot be overestimated, because before we can start calculating we must decide <u>what</u> to calculate.

Coherence can do the same for us in its more complex dynamic context. Even if it is no more than an idealization never strictly realized, it gives us a much better understanding of phenomena. To illustrate its predictive power, consider just one example: Let the flow of a multicomponent gas of constant composition through an adsorption column, with competitive adsorption equilibrium, be interrupted by a pulse of an inert gas; coherence theory without <u>any</u> calculation predicts the number and sequence of the resulting pulses (as many as there are sorbable components), the directions of the concentration variations in each, and the skewedness of each, apart from providing a complete picture of how and why this particular pattern arises (3).

Horizons

The concept of coherence was originally developed in 1963 for multicomponent fixed-bed ion exchange and adsorption operations (7), essentially preparative chromatography. Later major extension were to multiphase systems in enhanced oil recovery (5,8,10) and allowance for reactions at equilibrium (4,9,11-15) including precipitation (13-16) and micelle formation (11). A number of contributions to this symposium describe or review such developments.

But what can coherence do for us that it has not already done? What new and useful results can we still reasonably expect?

To answer this question, let us recall the gas absorption column, in which the concept of equilibrium contributed a much better understanding of what was happening and why, although bulk gas and liquid at any point were far from being in equilibrium with one another. It is almost certainly fair to say that in engineering practice the majority of applications of the equilibrium concept, even of thermodynamics more generally, are of this type: not a description of an equilibrium system, in which by definition "nothing happens," but an application to a system in a different state but with its tendency to approach equilibrium blocked or compensated by other forces.

The same can be said for the practical value of the coherence concept. For single, abrupt perturbations somewhere introduced and almost immediately resolved into coherent waves, coherence theory has little to contribute. It comes into its own in providing an understanding of the vastly more complex noncoherent situations - and simpler mathematics for handling them quantitatively. To date, the vast majority of applications of coherence theory has been to simple, completely or largely coherent systems. The power of the concept in the more general noncoherent context has as yet hardly been tapped.

As just one example of what I believe we can expect in the future, an investigation recently initiated in our laboratories comes to mind: a study of propagation of perturbations in countercurrent operations otherwise at steady state (17-19), with the long-range objective of providing a fundamental understanding of dynamic phenomena in multicomponent fractionation columns. We hope to come up eventually with guidelines for design of columns (or modification of existing ones) to make them less sensitive to perturbations and so operable at lesser reflux. With hundreds of thousands or even millions of such columns in operation, the potential savings in energy could be staggering.

The picture that is beginning to emerge is that of the column at steady state containing a number of standing waves, each corresponding to a different eigenvalue. A noncoherent perturbation at the feed tray is resolved into a set of response waves which are propagated up or down, each seeking out the standing wave of same eigenvalue, there to fade out eventually. As the response waves ride over the steady-state profile, their velocities change with the background compositions of the phases. Moreover, how sensitive any one standing wave is to perturbation depends strongly on whether its basic nature is self-sharpening or nonsharpening. An added complication is the possibility of wave reflection by the condenser and reboiler. The system is noncoherent wherever perturbed from to the steady state, the basic phenomenon being interference of the perturbation waves with the standing waves of the steady state. I believe we are beginning to understand the dynamic elements of the situation, but much work still remains to be done.

A perhaps even more challenging extension of coherence theory would be to reactors, for example, sparged towers, catalytic fixed beds, and especially trickle-flow reactors. Here, again, one must expect a perturbation to be resolved into response waves, which then travel over a steady-state composition profile. Added complications are the need to include chemical reactions with finite rates in the basic mass balances and, usually, to account for nonisothermal behavior, introducing temperature as an added dependent variable. However, both reactions (20) and nonisothermal behavior (21,22) have been included previously in the treatment of simpler, coherent systems by various other workers.

Lastly, we must realize that coherence theory is general in that it does not require time and distance to be the two independent variables. Systems with more dimensions in space, or with another independent variable instead of time, can be fitted into the general framework. Three-dimensional modeling of oil recovery operations or of chemical or oil spills and their clean-up - indeed, the entire field of what has recently come to be called chemodynamics - are obvious applications of the former. I would not even want to speculate at this point what doors a replacement of time by another variable might open.

Coherence in Broader Context

As this brief survey has shown, there is no lack of interesting problems to which coherence could be applied. However, it is not really the specifics of applications that matter most for an assessment of what impact coherence may have in the future.

I firmly believe we are at the threshold of a new era in industrial chemical engineering, an era with much greater emphasis on dynamics. When I left industry seven years ago, and I doubt much has changed since, design and optimization in grass-roots process development were done on a steady-state basis, and dynamics - that is, start-up, regular and emergency shut-down, controlability, etc. - was only invoked when the main features of the overall design were frozen. I know of at least two major recent processes that would look rather different, had they been optimized on a dynamic basis.

Today, our computers are large enough, fast enough, cheap enough for dynamic design and optimization. One bottleneck, in no relation to the topic at hand here, is software. The other, much related to it, is our still none too satisfactory fundamental understanding of dynamics of complex chemical engineering systems. A process offers an almost unlimited number of options, far too many for even the largest computer to evaluate. So, before proceeding to computerized optimization we have to make intelligent choices, based a good working knowledge of dynamics.

Of course, process dynamics is a well-established discipline, with skilled practitioners that know all the relevant equations, and with wonderful computers that can solve them. What more do we need? My point is that not equations, but their solutions are what tells the story, and that a computer run gives only the solution for that one specific set of input variables and conditions that was used. With systems as complex as is standard in chemical engineering practice, a room full of stacks of telephone books of output would have to be generated for a reasonably complete coverage, and to extract general rules and regularities from such a mass of paper becomes next to impossible. What more we need is a better grasp of cause and effect, an ability to foresee, to recognize opportunities and pitfalls, without need for extensive calculation and as a preparation for computerized quantitative optimization of conditions. Any concept that advances our understanding of fundamental process dynamics should prove of value in the time to come. I hope and trust coherence will be one such concept.

The Ultimate Challenge

The specific topics I mentioned a little earlier are opportunities which, I hope, will be picked up in the future. But there is one still greater and more fundamental challenge. It concerns the idea of coherence itself.

The coherence concept arose, twenty-five years ago, from a set of highly speculative and intuitive predictions I had casually made about the outcome of a major numerical study on multicomponent ion exchange columns planned by Vermeulen and Klein at the University of California, Berkeley. All these predictions were later borne out by the results of the first fifty or more cases solved numerically, to everyone's surprise including my own. What proved difficult, however, was to provide a rationale for the predictions: whatever explanation or trend of thought I tried soon became mired in contradictions or inconsistencies. It became clear to me that somehow I had glimpsed a greater truth without yet fully understanding it. The coherence concept then provided the key, even though its proof proved more elusive.

What still both fascinates and frustrates me about coherence is that even today I do not fully understand it - nor does, I believe, anyone else. While certainly not valid without exceptions (after all, neither are equilibrium and steady state), the concept appears to be much more general than any of its proofs so far provided. Specifically, the rigorous proof of attainment of coherence from arbitrary starting conditions (6), with use of the method of characteristics, is restricted to hyperbolic differential equations. If simplifying premises are relaxed, the equations become parabolic and that proof no longer holds. However, it can be shown laboriously and indirectly that the second-order effects now admitted either "die out" or establish a time-independent balance, so that coherence nevertheless is asymptotically approached. In more general terms, it seems that whatever new type of systems one turns to, attainment of (or development toward) coherence must be proved anew, for the different type of fundamental equations. It seems likely that a greater truth is still buried somewhere, that it should be possible to give a proof of development toward coherence in a more general form, independent of the specific nature of the equations of the system. Moreover, such proof would help to delineate when coherence may or may not apply.

Incidentally, such a proof - a "general coherence theory," if you will - would also apply to equilibrium (not to mention steady state), whose attainment we have come to accept unquestioningly. Indeed, one of the greatest wonders in the development of our scientific thinking has been thermodynamics, a fantastic edifice built entirely on a few unproved basic postulates. Today, any referee or editor would reject the writings of Gibbs, Clausius & Co. as unsubstantiated speculation. But for a hundred years thermodynamics has been successful without exception, and we have all but forgotten to ask for proof, have become preoccupied by it to the point of applying its equilibrium formalism to rate phenomena, as in transition-state theory and thermodynamics of irreversible processes. But unless we are able to overcome this preoccupation with thermodynamics, Gibbs' genius will ultimately have done us a disservice. In the science of all other fields of physics and engineering - mechanical, electrical, etc. - the principal role is played by dynamics, not statics. Our physical chemistry and chemical engineering are the only fields dominated by a science of equilibrium, of statics (for thermo-"dynamics" is, of course, a classic misnomer). I believe the time has come for us to break with this domination, to pay more attention to dynamics in our field, so far the poor step-sister of thermodynamics. A general coherence theory could be a step in this direction and would greatly advance our understanding of the world we live in. I do hope and dream that some day in the future a mind better than mine will pick up this ultimate challenge and will succeed where I have failed.

Acknowledgements

The coherence concepts owes much to two persons who at the time of this writing are no longer with us: Gerhard Klein, of the University of California, Berkeley, whose sharp analytical mind did much in the early days to bring quantitative structure into my conceptual and intuitive thinking, and my wife, whose patience allowed me to invest so much time in what is being honored here. The Award in Separation Science and Technology, matched two dollars for one by the Shell Foundation, has helped to build a scholarship fund in her memory.

Literature Cited

1. Helfferich, F. G. AIChE Symp. Ser. 1984, 80, 1-13.
2. Helfferich, F. in Adsorption From Aqueous Solution;
   American Chemical Society, Washington, DC, 1968, p 37.
3. Helfferich, F. G.; Klein, G. Multicomponent Chromatography;
   Marcel Dekker: New York, 1970, Chapters 3, 4 (out of print,
   obtainable from University Microfilms International, Ann
   Arbor, Michigan, #2050382).
4. Helfferich, F. G.; Bennett, B. J. Solvent Extr. & Ion
   Exchange 1984, 2, 1151-1184.
5. Helfferich, F. G. SPE J. 1981, 21, 51-62.
6. Helfferich, F. G. Chem. Eng. Comm. 1986, 44, 275-285.
7. Helfferich, F. G. Theory of Chromatography: Basis of a
   Generalized Theory for Multicomponent Systems With
   Interdependent Nonlinear Isotherms; Shell Development
   Company, Technical Progress Report 315-64, 1964.
8. Hirasaki, G. J. SPE J. 1981, 21, 191-204.
9. Klein, G. in Percolation Processes, Theory and
   Applications; Rodrigues, A. E.; Tondeur, D., Eds.; Sijthoff
   and Noordhoff, Alphen aan den Rijn, 1981, pp 363-421.
10. Hirasaki, G. J. SPE J. 1982, 22, 181-192.
11. Harwell, J. H.; Helfferich, F. G.; Schechter, R. S. AIChE
    J. 1982, 28, 448-459.
    Helfferich, F. G.; Bennett, B. J. React. Polymers 1984, 3,
12. 51-66.
13. Walsh, M. P.; Bryant, S. L.; Schechter, R. S.; Lake, L. W.
    AIChE J. 1984, 30, 317-328.
14. Bryant, S. L.; Schechter, R. S.; Lake, L. W. AIChE J. 1986,
    32, 751-764.
15. Bryant, S. L.; Schechter, R. S.; Lake, L. W. AIChE J. 1987,
    33, 1271-1287.
16. Klein, G. in Ion Exchange: Science and Technology;
    Rodrigues, A. E., Ed.; Nijhoff, Dordrecht, 1986, pp
    199-226.
17. Hwang, Y.-L. Ph.D. Thesis, Pennsylvania State University,
    University Park, PA, 1987.
18. Hwang, Y.-L. Chem. Eng. Sci. 1987, 42, 105-123.
19. Hwang, Y.-L.; Helfferich F. G. Dynamics of Continuous
    Countercurrent Mass-Transfer Processes. II. Chem. Eng.
    Sci. 1988, 43, in press.
    Schweich, D.; Villermaux, J.; Sardin, M. AIChE J. 1980, 26,
20. 477-486.
    Rhee, N.-K.; Amundson, N. R. Chem. Eng. J. 1970, 1,
21. 241-254.
22. Rhee, H.-K.; Heerdt, E. D.; Amundson, N. R. Chem. Eng. J.
    1970, 1, 279-290.

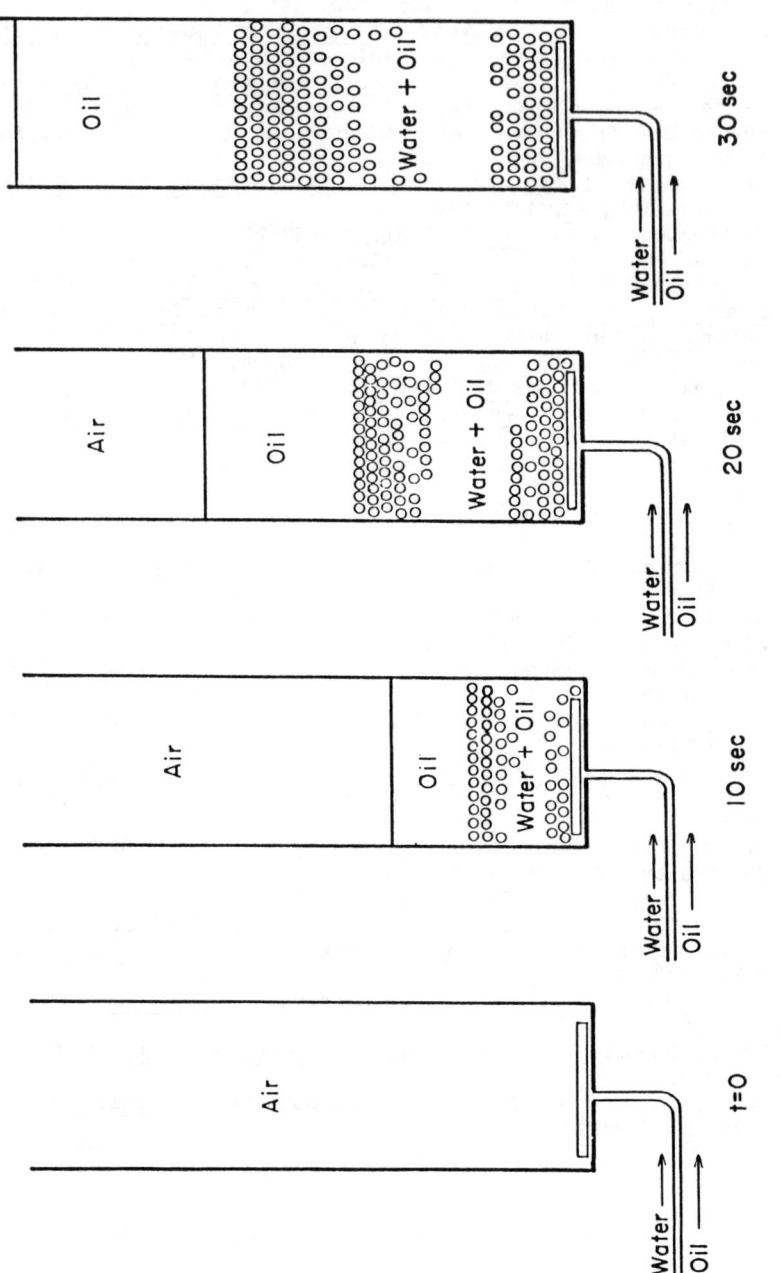

Figure 1. A simple model system: injection of water-oil mixture into open tube. (Reproduced with permission from Ref. 1. Copyright 1984 American Institute of Chemical Engineers.)

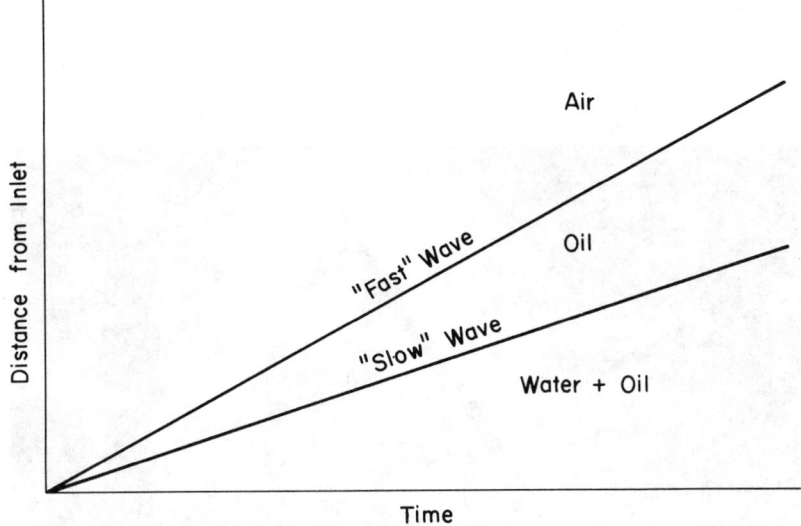

Figure 2.   Distance-time diagram for system in Figure 1. (Adapted with permission from Ref. 1.   Copyright 1984 American Institute of Chemical Engineers.)

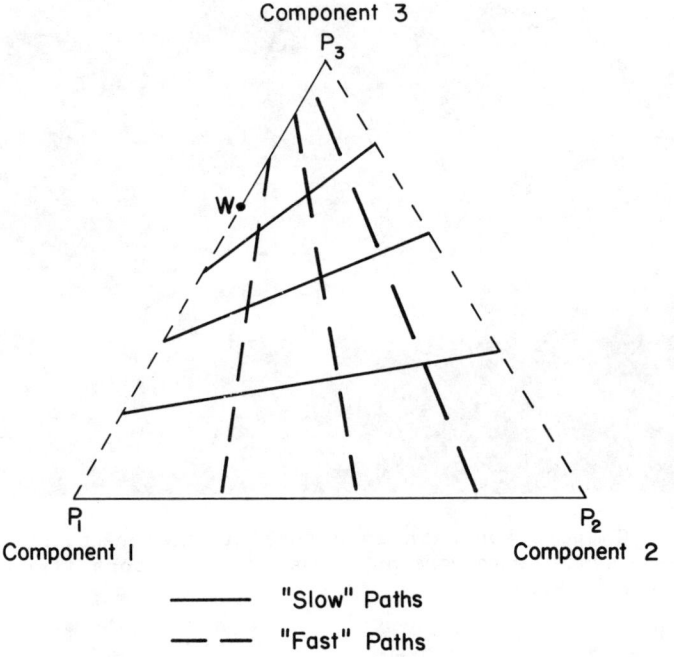

Figure 3.   Composition path grid for three-component, fixed-bed ion exchange system with constant separation factors (adapted from Ref. 2, see also Ref. 3).

Figure 4. Composition path grid for four-component, fixed-bed ion exchange system with constant separation factors (from Ref. 2, see also Ref. 3).

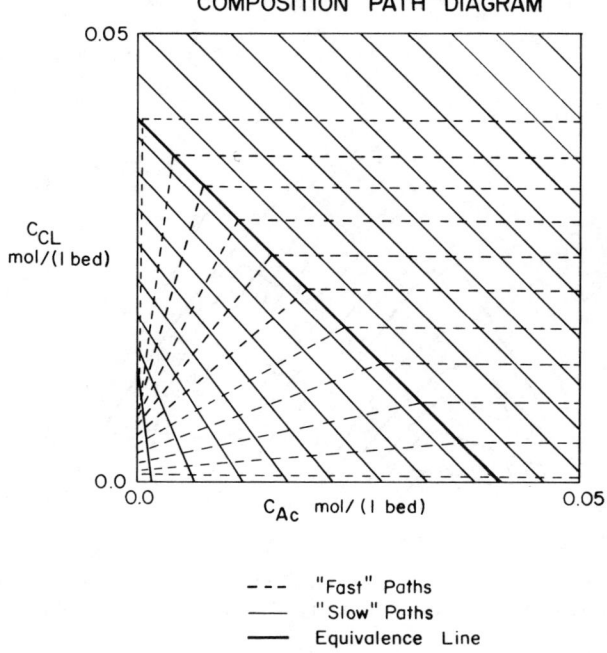

Figure 5. Composition path grid for acetate-chloride fixed-bed anion exchange system with constant separation factors and dissociation equilibrium (adapted with permission from Ref. 4. Copyright 1984 Marcel Dekker).

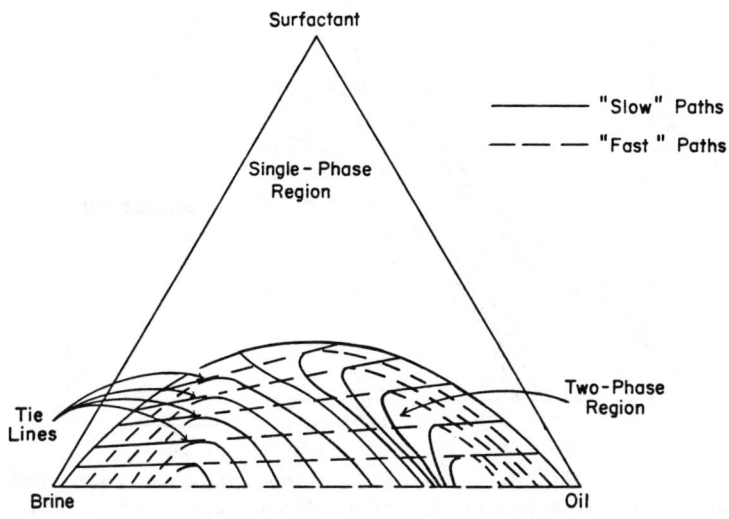

Figure 6. Composition path grid for brine-oil-surfactant system in porous medium (adapted with permission from Ref. 5. Copyright 1981 Society of Petroleum Engineers).

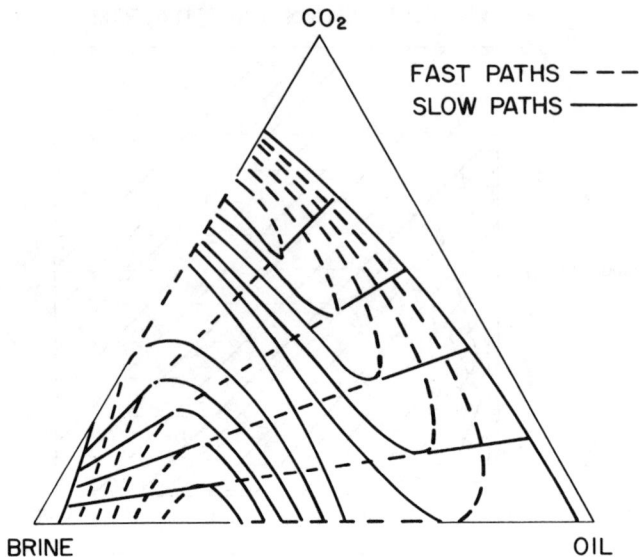

Figure 7. Composition path grid for brine-oil-carbon dioxide system in porous medium (Helfferich, F. G.; Hirasaki, G. J. 1981, unpublished).

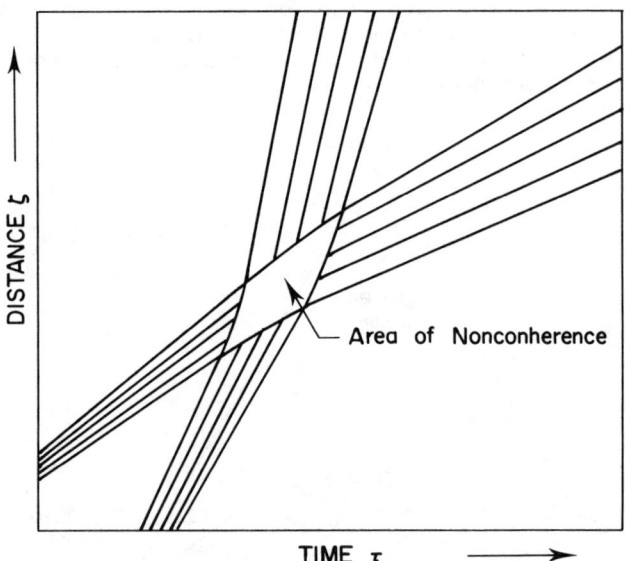

Figure 8. Interference of diffuse coherent waves in distance-time plane. Curves represent trajectories of given compositions. (Adapted with permission from Ref. 1. Copyright 1984 American Institute of Chemical Engineers).

Chapter 2

An Overview of Coherence in the Chromatographic Movement of Surfactant Mixtures

Jeffrey H. Harwell

This paper summarizes results from the application of coherence theory to the chromatographic movement of surfactant mixtures. Emphasis is placed upon the ability of coherence theory to produce generalizations concerning the behavior of a system which includes highly nonideal, nonlinear elements. Interesting and unexpected results are obtained, including the formation of waves at the Critical Micelle Concentration of the mixture which propagate at the velocity of a non-adsorbing tracer.

The very aspect of surfactant behavior that makes the class of molecules interesting, their surface activity, makes for some apparently complicated behavior in solution, and for some even more complicated behavior in the chromatographic behavior of a mixture of surfactants. By thinking of the chromatographic behavior in terms of the formation of coherent waves, however, generalizations result which enable one to make predictions about the behavior of a system without being distracted by the large number of details which must be accounted for in a rigorous mathematical model. When considered in light of the tendency of the system to resolve itself into coherent waves, the gross behavior reduces to the interaction of a few readily understood parameters.

Surfactant Solution Behavior

Micelle Formation. To have sufficient surface activity to be considered a surfactant, a molecule must have hydrophobic and hydrophilic moieties

spatially related in such a way as to allow the
hydrophilic moiety to be in contact with the aqueous
phase while the hydrophobic moiety is removed from the
aqueous phase when the molecule goes to an interface.
It is found that surfactants will aggregate in
solution to form micelles, which may usefully be
thought of as aggregates of 50-200 molecules, oriented
so that the hydrophilic moieties form a hydrophilic
coat around the aggregated hydrophobic moieties; one
will not go too far wrong in understanding the
properties of a micelle if it is thought of as a
droplet of liquid hydrocarbon inside a porous
hydrophilic shell.   The droplet of hydrocarbon is made
up of the hydrocarbon tails of the surfactants in the
micelle, and the shell is made up of their hydrophilic
head groups.   Surfactant solutions do not form
micelles at all concentrations, but only at
concentrations higher than a certain critical
concentration, referred to as the Critical Micelle
Concentration (CMC).
     Micelle formation dramatically affects the
properties of surfactant solutions.   Lowering of
interfacial tension is an important example.   Below
the CMC surface tension decreases with increasing
surfactant concentration, because the surfactant is
adsorbing at the surface.   Once the CMC is reached,
the surface tension stops changing even when more
surfactant is added to the solution.   A related
phenomenon occurs in surfactant adsorption at a
solid/solution interface.   As would be expected, below
the CMC adsorption increases proportionally to
surfactant concentration increases.   After the CMC is
reached, however, the micelles act as a chemical
potential sink for the surfactant, and additional
surfactant forms more micelles without increasing the
adsorption significantly.   In these situations the
chemical potential of the surfactant depends only on
the concentration of the unaggregated or monomeric
surfactant.   This is so much like a phase change that
models of the phenomenon are referred to as pseudo
phase separation models (1,2).

Mixed Micelle Formation.   If two or more surfactant
species are present, the situation is further
complicated by the differential distribution of the
surfactants between the monomer pseudo-phase and the
micelle pseudo-phase.   The phenomenon is analogous to
the cooling a mixture of hydrocarbon gases to the dew
point of the mixture; when the first droplet of liquid
appears, we do not expect it to have the same mole
fraction distribution of hydrocarbons as the gas
mixture.   Instead, we expect the first droplet of
liquid to be enriched in the less volatile components
of the mixture.   In the same way, the micelles will

be enriched in the molecules with the lowest CMCs.
This behavior can be modeled by approaches analogous
to ideal solution or regular solution vapor/liquid
equilibria (3-5).

When a model for surfactant monomer/micelle
equilibrium is combined with a model which predicts
the adsorption of the surfactant from its monomer
concentration, the adsorption of the surfactant
components can be predicted for the mixture (6,7).

Effect of Mixed Micelles on Chromatography

In order to separate out the effects of micelle
formation on the chromatographic movement of
surfactant mixtures from any possible effects due to
surfactant/surfactant interactions at the solid
surface, we first consider a hypothetical, binary
surfactant system in which each surfactant's
adsorption is given by a Henry's law coefficient
multiplied times the concentration of its monomer. We
also assume that the monomer/micelle equilibrium is
adequately described by ideal mixed micelle formation
(8).

Constructing the Composition Path Grid. The first
step to understanding the effects of mixed micelles on
surfactant chromatographic movement is to construct
the composition path grid for the system (9). So
far, the mathematics of coherence have only been
worked out for systems described by hyperbolic
equations, so typical coherent paths for the system
can only be constructed by neglecting dispersion in
the material balance equations of the surfactants:

$$\frac{\partial C_{i,T}}{\partial t} + V \frac{\partial C_i}{\partial z} = 0 \quad i = 1,2$$

where

$$C_{i,T} = \overline{C}_i + C_i$$
$$C_i = \phi c_i$$
$$C_{i,d} = \phi c_{i,d}$$ (1)
$$C_{i,m} = \phi c_{i,m}$$
$$\overline{C}_i = (1 - \phi)\hat{\rho}_s q_i$$

We do not need to specify initial and boundary
conditions in order to construct the composition path
grid.

A coherent wave is a propagationally stable
variation in the concentration of components in the
system. The general expression for this is as
follows:

$$V_{c_1} = V_{c_2} = \ldots = V_{c_n}$$

where

$$V_{c_i} = (\partial z / \partial t)_{c_i} \qquad i = 1,2,\ldots,n \tag{2}$$

An equivalent expression for the concentration velocity can be written:

$$V_{c_i} = -\left(\frac{\partial C_{i,T}}{\partial t}\right)_z \bigg/ \left(\frac{\partial C_{i,T}}{\partial z}\right)_t = -\left(\frac{\partial C_i}{\partial t}\right)_z \bigg/ \left(\frac{\partial C_i}{\partial z}\right)_t \tag{3}$$

Substituting the material balances into this expression, and noting that all the concentration velocities must be equal in a coherent wave, we arrive at this relationship:

$$\left(\frac{\partial C_{i,T}}{\partial z}\right)_t = +\frac{1}{\lambda}\left(\frac{\partial C_i}{\partial z}\right)_t$$

$$\left(\frac{\partial C_{i,T}}{\partial t}\right)_z = \frac{1}{\lambda}\left(\frac{\partial C_i}{\partial t}\right)_z \tag{4}$$

$$\lambda = \frac{V_{ci}}{V} = \frac{V_{cj}}{V} \qquad \text{for } i = 1,\ldots n$$

We can consider the dependent variable $C_{i,T}$ as a function of either the solution concentration, C_i, or of the independent variables in the material balance; thus, the next two expressions must also be used:

$$dC_{i,T} = \sum_{j=1}^{n} \frac{\partial C_{i,T}}{\partial C_j} dC_j = \sum_{j=1}^{n} m_{i,j}\, dC_j \tag{5}$$

$$dC_{i,T} = \left(\frac{\partial C_{i,T}}{\partial z}\right)_t dz + \left(\frac{\partial C_{i,T}}{\partial t}\right)_z dt \tag{6}$$

These expressions allow us to rewrite the coherence condition in terms of the dependent variables only:

$$dC_{i,T} = \frac{1}{\lambda}\left(\frac{\partial C_i}{\partial z}\right)_t dz + \frac{1}{\lambda}\left(\frac{\partial C_i}{\partial t}\right)_z dt = \frac{1}{\lambda} dC_i$$

$$dC_i = \lambda \sum_{j=1}^{n} m_{i,j} dC_i \tag{7}$$

The matrix form of this expression for a binary system is as follows:

$$\begin{bmatrix} \lambda m_{1,1} - 1 & \lambda m_{1,2} \\ \lambda m_{2,1} & \lambda m_{2,2} - 1 \end{bmatrix} \begin{bmatrix} dC_1 \\ dC_2 \end{bmatrix} = \begin{bmatrix} 0 \\ 0 \end{bmatrix} \tag{8}$$

The characteristic equation from this matrix is

$$(\lambda m_{1,1} - 1)(\lambda m_{2,2} - 1) - m_{1,2} m_{2,1} \lambda^2 = 0 \tag{9}$$

This gives rise to expressions for the eigenvectors and the eigenvalues which satisfy the coherence condition in the space of the dependent variables:

$$\lambda = \frac{m_{1,1} \pm ((m_{1,1} - m_{2,2})^2 + 4m_{1,2} m_{2,1})^{1/2}}{2(m_{1,1} m_{2,2} - m_{1,2} m_{2,1})} \tag{10}$$

$$\frac{dC_1}{dC_2} = \frac{\lambda m_{1,2}}{1 - \lambda m_{1,1}} \tag{11}$$

In order to construct one composition path in the (C_1-C_2) plane, one selects a point in the plane near one of the axes, then integrates equations 10 and 11 out into the plane. At each point on the curve the expressions describing the distribution of the total concentration of each component, the adsorption, and the monomer-micelle equilibrium must all be satisfied:

$$C_{i,t} = \overline{C}_i + C_i$$
$$\overline{C}_i = K_i C_{i,d}$$

$$\phi \sum C_{i,d} = \left[\frac{(\Pi \phi C_i^*)^\theta}{\sum x_i \left(\prod_{i \neq j} \phi C_j^* \right)^\theta} \right]^{1/\theta} \tag{12}$$

$$\frac{C_{1,m}}{C_{2,m}} = \frac{C_{1,d}}{C_{2,d}} \left(\frac{C_2^*}{C_1^*}\right)^\theta$$

The easiest way to do so is to numerically integrate
the expression describing the eigenvectors. This
procedure must be followed for each root of the
expression for the eigenvalues.

A composition path grid constructed by this
procedure for one particular hypothetical system is
shown in Figure 1. The dominant feature of the grid
is the curve which represents the CMC of the
surfactant mixture. Above the CMC there are mixed
micelles in solution. Below the CMC no micelles
exist. From each point on the CMC two families of
coherent paths (eigenvectors) radiate out above the
CMC, and two families of coherent paths radiate out
below the CMC. The most surprising feature of the
composition path grid is that the eigenvalues of the
concentrations along the family of straight lines
radiating out from the CMC are all unity. This
implies that coherent waves can be formed at
compositions above the CMC which move at the velocity
of a non-adsorbing component, a tracer, for example.
If we examine the values of the adsorption of the
surfactants along one of these lines, we find that
they are also lines of constant adsorption, constant
monomer concentrations, and constant micelle
composition. All that varies along one of these
curves is the concentration of micelles. In other
words, if I locate two points on one of the straight
lines radiating out above the CMC, the two points
represent two surfactant concentrations that would be
in equilibrium with the same micelles and the same
adsorbed concentration. This explains how a coherent
wave of variation in surfactant concentration could
propagate at the velocity of a tracer.

In order to determine the coherent waves the
system will tend to form for any specific initial and
boundary conditions, basically all that is necessary
is to locate the initial and boundary conditions on
the composition path grid, then connect them by a
route that always follows one of the coherent paths.
If we begin with the boundary condition, then we must
follow the path passing through the point representing
the boundary condition which has the lower
eigenvelocity. We continue on this path until it
intersects the path with the higher eigenvelocity
which passes through the point representing the
initial conditions. (There are some slight
complications to this procedure, but they are easily
applied, and situations in which they apply are easily
recognized.)

To illustrate how this works, let us consider two
concentrations below the CMC. In Figure 2a the
concentration labelled O is the concentration of
surfactant in a chromatographic column at time zero;
the concentration labelled I is the concentration

which is injected into the column beginning at time
zero.   After some finite volume of composition I is
injected, we begin injecting concentration O again.
First, we follow the path with the lower eigenvalue
from I to B, where it intersects the path passing
through O which has the higher eigenvelocity.

Composition B is a composition which will arise within
the system as the original non-coherent variation
between I and O is resolved by the system into the
propagation of two coherent waves.   When we consider
the new variation between O and I which results from
our injection of composition O behind the pulse of
surfactant, we see that it will be resolved into a
variation between O and C and a variation between C
and I; we use the same rules to determine which path
to follow going out from O.

The trajectories of the wave of variation I/B and
the wave of variation B/O can be plotted (Figure 2b),
along with the trajectories of the wave of variation
O/C and of the wave of variation C/O, in a normalized
distance (pore volumes travelled) versus normalized
time (pore volumes injected) diagram; it is observed
that the C/O wave eventually overtakes the I/O wave.
When this happens, the composition I has disappeared
from the column; a new non-coherence has arisen,
however: the variation between C and B.   The system
resolves this non-coherence also along coherent paths,
as shown by Figure 2a.

Concentration profiles for the column at various
dimensionless times can be constructed from the plots
of the wave trajectories, as shown in Figure 2c.
Having done this we observe that our example is a
rather trivial one, the propagation of concentrations
of two non-interacting solutes.   Each one propagates
independently of the other.   With a simple Henry's Law
adsorption behavior for each component, the injection
of a pulse of the two components results in the
propagation of two square waves down the column.

Injection of a composition above the CMC allows
us to observe the effects of mixed micelle formation
on the propagation of the surfactants.   In Figure 3a
the initial conditions are represented by point "O"
and the boundary conditions by point "I".   We would
normally expect to resolve the non-coherence between O
and I along composition paths; however, we find that
the composition velocities along the slow path passing
through I are decreasing in the direction we must
follow to reach a fast path passing through O.   When
this situation arises, downstream concentrations have
a lower velocity than the upstream concentrations, and
a shock wave is formed.   The continuum assumption used
to formulate the material balance equations is no
longer valid in this case, and the velocity of the
wave between I and B, as well as the composition B

itself, must be found by trial and error using a mass
balance across the shock wave:

$$V_{c_i} = \frac{V}{1 + \dfrac{\Delta \overline{C}_i}{\Delta C_i}}$$

$$V_{c_i} = V_{c_2}$$

Injection of the composition O behind the
surfactant pulse of composition I produces another
non-coherence, which is resolved into the variations
O/C, C/D, and D/I. The trajectories of these
variations are all plotted in Figure 3b. We make
three observations of particular interest from Figure
3b. Note that there are now three waves generated
behind the surfactant pulse, that the wave of
variation D/I propagates at a normalized velocity of
1, and that composition D is precisely on the CMC of
the system. When the distance/time diagram is used to
construct column profiles, shown if Figure 3c, we see
that wave D/I sweeps rapidly up the back of the pulse;
when in interferes with wave I/B at the front of the
pulse, there are no micelles left in the column.
The physics of what has happened are quite
interesting. Composition D is on the same composition
path as composition I, one of the composition paths
along which the eigenvelocity is unity. The micelles
at composition D, and all the compositions between I
and D, are already in equilibrium with the monomer
concentration, and thus with the adsorbed phase, downstream at composition I. These micelles thus sweep
along with the bulk fluid phase, undergoing no
variation in composition due to adsorption or
chromatographic separation. It is especially to be
observed that micelles of this composition were not
injected into the column--they formed spontaneously as
the system resolved itself into coherent waves!

Experimental Verification of the Effects of Mixed Micelles

There is no real system which corresponds exactly to
the system whose behavior was briefly examined above.
Experimental verification of the spontaneous formation
of waves at the system CMC propagating at a
normalized velocity of 1 requires the introduction of
such inescapable real-world complications as
dispersion. For the particular system selected for
study, there are also effects arising from the
formation of mixed admicelles, adsorbed aggregates of

surfactants which are much like mixed micelles. While
we can no longer quantitatively predict the behavior
of the system (the introduction of dispersion into the
mass balance equations makes them parabolic),
coherence theory still predicts that the system will
settle into propagationally stable waves. To find
these waves, we use the same model applied above to
the hypothetical system; the only modification is the
use of a realistic adsorption model capable of
describing the formation of mixed admicelles (10).
    The composition path grid for the real system,
shown in Figure 4, is constructed using the procedure
described above for the hypothetical system. It is
based only upon the pure component adsorption data and
the pure component CMC data. No mixture experiments,
and no chromatographic experiments were performed in
order to construct the path grid for the system. As
in the composition path grid for the hypothetical
system, the straight lines radiating out from the
mixture CMC curve are lines of constant adsorption,
and thus have an eigenvelocity of 1 for every point
along them.

Injection of a Solution Enriched in the Less Strongly
Adsorbing Surfactant. To illustrate the effects of
mixed micelles on the propagation of the surfactant
mixture, we inject a composition located at point I in
Figure 5a into a column equilibrated with the
composition at point O. Using the same principals as
before, we predict a resolution of the non-coherence
between I and O into two coherent waves, one of
composition variation I/A, the other of variation A/O.
The wave of variation I/A is a spreading wave; the
eigenvelocities increase in the downstream direction.
When the downstream concentrations have greater
velocities than the upstream concentrations, the wave
spreads out as it propagates. In this case, however,
the difference in the eigenvelocities along the
composition path is very small, as can be seen from
the corresponding distance/time diagram, Figure 5b.
The wave of variation A/O is along a unit velocity
path; it is predicted to be a variation in micelle
concentration, at constant micelle composition and at
constant adsorption, which will propagate through the
column at the velocity of a non-adsorbing tracer.
    Notice also that while the concentration of
surfactant 1 is doubled in the injected solution, the
concentration of surfactant 2 is the same in both the
original and the injected solution. Nevertheless, we
predict that the intermediate composition which will
arise between I and O will have a doubled
concentration of the surfactant 2. This doubling of
the concentration of the surfactant 2 will occur

because micelles in solution O are enriched in
surfactant 2 relative to the micelles in solution I.

The observed and predicted compositions are shown
in Figure 5c; observed and predicted effluent profiles
are shown in Figure 5d. While there is much of
interest to discuss regarding these figures, we
confine ourselves to observing that a broad plateau is
formed in the effluent which is, as predicted, at
composition A, where the concentration of the more
strongly adsorbed component is doubled over the
concentration in either the original or the injected
solutions. Even more striking, if we calculate
the velocity of the wave of variation A/O from the
midpoint between the plateau at O and the plateau at
A, we obtain a value of the normalized wave velocity
of 0.96. For the normalized wave velocity to have
been 0.96, the adsorption of surfactant 2 can have
varied by only 0.5% between A and O, though the
concentration of surfactant 2 varied by 100%. To
understand this phenomenon, imagine the composition at
A to have the same monomer composition as the
composition at O, but with approximately twice as many
micelles in equilibrium with that monomer. Since
there is no adsorption from the micelles at A as they
propagate down the column, they travel at the velocity
of a tracer.

One naturally has to ask, after looking at Figure
5d, why the velocity of wave A/O is so accurate, why
the plateau compositions are so accurate, and yet the
velocity of wave I/A is off by nearly 50%. The
explanation is in the relative adequacy of the models
of the physical phenomena that determine the plateau
concentrations and of the wave velocities. The
velocity of wave A/O is completely determined, and the
plateau value is largely determined, by the formation
of mixed micelles. Ideal mixing very accurately
describes the monomer/micelle equilibrium for this
system, so the velocity of A/O and the plateau
concentration are also very accurate. The velocity of
wave I/A, however, primarily depends upon the
variation in the adsorption of both surfactants
between compositions O and I (the adsorption at A is
the same as the adsorption at O). The calculations of
the composition path grid were performed using a
multicomponent surfactant adsorption isotherm based on
only the pure component adsorption data (10). The
isotherm is accurate enough to adequately predict the
plateau composition (A) at the concentrations used;
nearer the CMC, where fewer micelles are present, the
ability of the isotherm to predict the plateau should
decrease. The velocity calculation for wave I/A tends
to magnify the inadequacy of the isotherm, so the
predicted velocity of I/A is less accurate.

<u>Injection of a Solution Depleted in the Less Strongly Adsorbing Surfactant.</u> To further illustrate the effects of mixed micelles on the propagation of the surfactant mixture, we inject a composition located at point I in Figure 6a into a column equilibrated with a solution at point O in Figure 6a. Composition I has only half the concentration of the less strongly adsorbing surfactant, surfactant 1, than composition O, but has the same concentration of surfactant 2. Resolution of the non-coherence between I and O gives rise to coherent variations I/A and A/O. The wave of variation A/O is predicted to have a normalized velocity of 1. Also, the solution at A is predicted to have only half the concentration of surfactant 2 found in either the original or the injected solutions.

The observed effluent compositions are shown in Figure 6c, and can be seen to correspond very closely to the predicted compositions. The observed and predicted effluent profiles are shown in Figure 6d. We again note the spontaneous formation of micelles which are in equilibrium with the downstream surfactant concentrations. Calculating the velocity of the wave of variation A/O as was done in the previous example, we obtain a normalized velocity of 0.98. For the wave to have propagated at this velocity the adsorption of surfactant 2, the more strongly adsorbing surfactant, could have varied by only 0.2% from A to O, even though the concentration of surfactant 2 is doubled in O relative to A.

Conclusions

The formation of mixed micelles gives rise to some unusual and unexpected chromatographic behavior. It would, of course, be possible to model the effluent profiles in Figures 5 and 6 with a model which includes dispersion. The value of the dispersion coefficient could be predicted from correlations, or obtained from a best fit of the profiles. Without, however, the composition path grids constructed from coherence theory, how long would it have taken to obtain the generalizations about the propagation of surfactant mixtures for which coherence theory looks? It is easily conceivable that one could have run twice as many experiments, have performed twice as many numerical simulations, and have used ten times the computer time, without having ever recognized that the unit velocity waves were waves of variation in micelle concentration, at constant micelle composition.

Acknowledgments

This work was performed as part of the author´s dissertation work at the University of Texas. Dr. R.S. Schechter, Department of Chemical Engineering and Department of Petroleum Engineering, was the supervising professor.

Legend of Symbols

c_i = moles of i in fluid phase/m^3 of fluid
C_i = moles of i in fluid phase/m^3 of column ($C_i = \phi c_i$)
$C_{i,d}$ = moles of i as monomer (disperse phase)/m^3 of column
$C_{i,in}$ = surfactant concentration at inlet
$C_{i,m}$ = moles of i in micellar form/m^3 of column
$C_{i,o}$ = initial surfactant concentration
$C_{i,t}$ = total concentration of i, moles of i/m^3 of column
$\quad$ ($C_{i,t} = C_i + \overline{C}_i = C_{i,m} + C_{i,d} + \overline{C}_i$)
$\overline{C}_i$ = moles of i adsorbed/m^3 of column ($\overline{C}_i = (1 - \phi)q_i/\hat{\rho}_s$
c_i^* = pure component CMC in absence of added electrolyte (moles of i/m^3 of fluid)
CMC = concentration of surfactant at which micelles begin to form according to the pseudophase separation model of Mysels and Otter
$\Delta \overline{C}_i$ = [$\overline{C}_i$ behind a shock] − [$\overline{C}_i$ ahead of the shock]
ΔC_i = [C_i behind a shock] − [C_i ahead of the shock]
k_i = Henry's Law constant (moles of i adsorbed per kg of substrate)/(mole of i per m^3 of fluid)
K_i = Henry's Law constant (dimensionless) $K_i = k_i \rho_s (1 - \phi)/\phi$
q_i = adsorbed concentration of component i (moles of i adsorbed/kg of substrate)

Greek Letters

$\alpha_{i,j}$ = ≡ $[C_i^*/C_j^*]^\theta$
θ = empirical constant used to predict multicomponent CMCs (dimensionless)
λ = eigenvelocity of a wave defined by Eq. 12
ξ = dimensionless distance, (pore volumes from injection point)/(slug volume)
$\hat{\rho}_s$ = solid phase density (kg/m^3)
τ = dimensionless time, (pore volumes injected)/(slug volume)
ϕ = column void fraction (m^3 of fluid phase)/(m^3 of column)

Literature Cited

1. Shinoda, K. *Colloidal Surfactants*; Academic Press, New York, 1963, p 65.
2. Harwell, J.H.; Hoskins, J.; Schechter, R.S.; Wade, W.H. *Langmuir* 1985 **1**, 251.
3. Mysels, K.J.; Otter, R.J. *J. Colloid Sci.* 1961, **16**, 474.
4. Clint, J.H. *J. Chem. Soc., Faraday Trans. I* 1975, **71**, 1327.

5. Rubingh, D.N. <u>ACS Colloid Surface Sci Symp., 52nd</u>
    <u>(Knoxville)</u> 1978, <u>1</u>, 185.
6. Scamehorn, J.F.; Schechter, R.S.; Wade, W.H. <u>J.</u>
    <u>Colloid Interface Sci.</u> 1982, <u>85</u>, 463.
7. Trogus, F.J.; Schechter, R.S.; Wade, W.H. <u>J.</u>
    <u>Colloid Interface Sci.</u> 1979, <u>70</u>, 293.
8. Harwell, J.H.; Helfferich, F.G.; Schechter, R.S.
    <u>AICHE J.</u> 1982, <u>28</u>, 448.
9. Helfferich, F.G.; Klein, G. <u>Multicomponent</u>
    <u>Chromatography: Theory of Interference</u>; Marcel
    Dekker, New York, 1970.
10. Harwell, J.H.; Schechter, R.S.; Wade, W.H.
    <u>AICHE J.</u> 1985, <u>31</u>, 415.

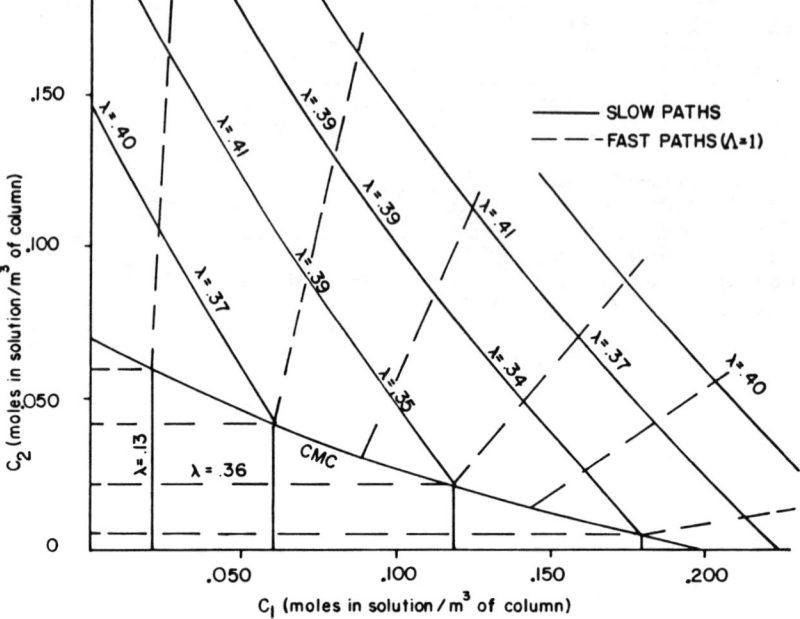

Figure 1.   Composition path grid for a hypothetical ideal mixed surfactant system.   (Reproduced with permission from Ref. 8, copyright 1982, American Institute of Chemical Engineers.)

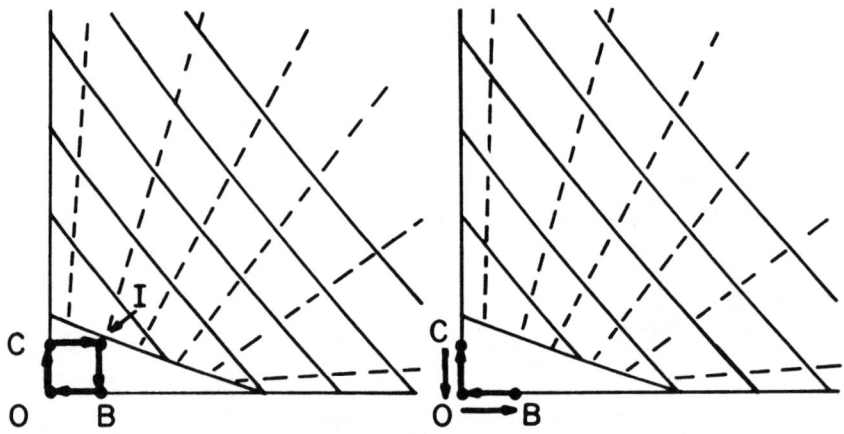

Figure 2.   Chromatographic behavior of an ideal, hypothetical mixed surfactant system in the absence of micelles.
(a) Composition route development   (b) Distance-time diagram   (c) Column profiles.   (Reproduced with permission from Ref. 8, copyright 1982, American Institute of Chemical Engineers.)

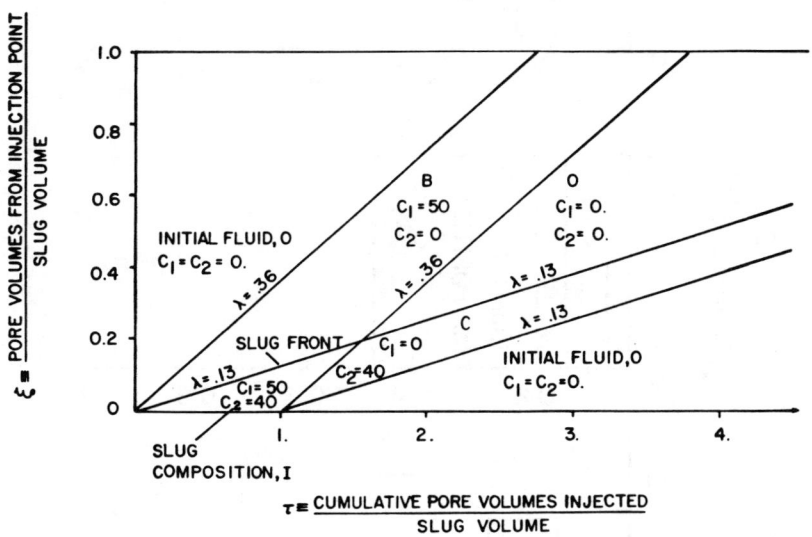

Figure 2B.

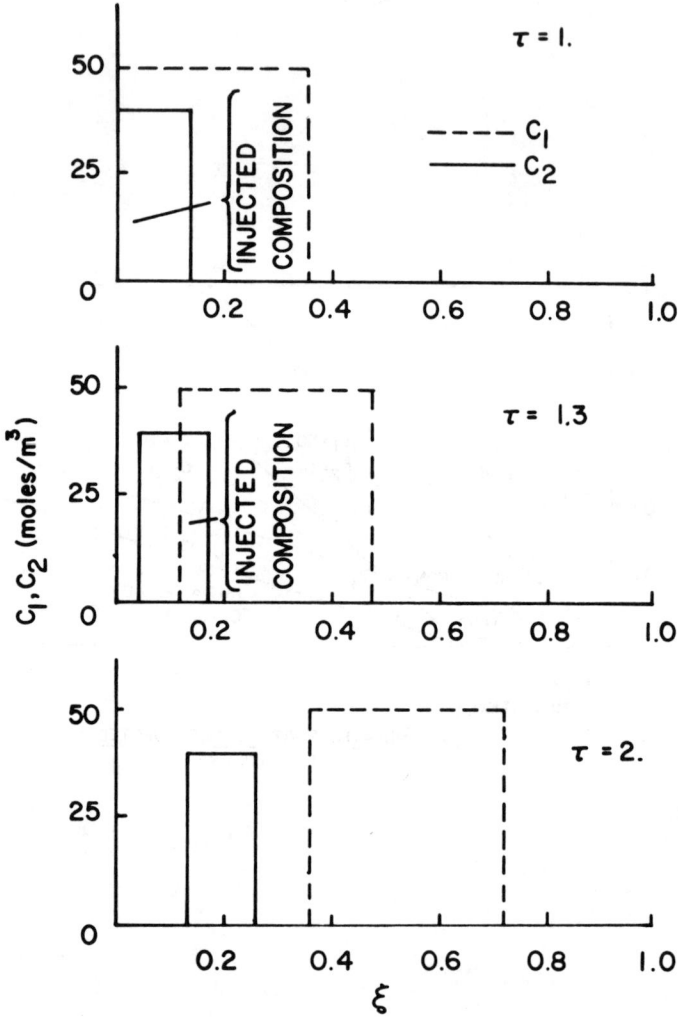

Figure 2C.

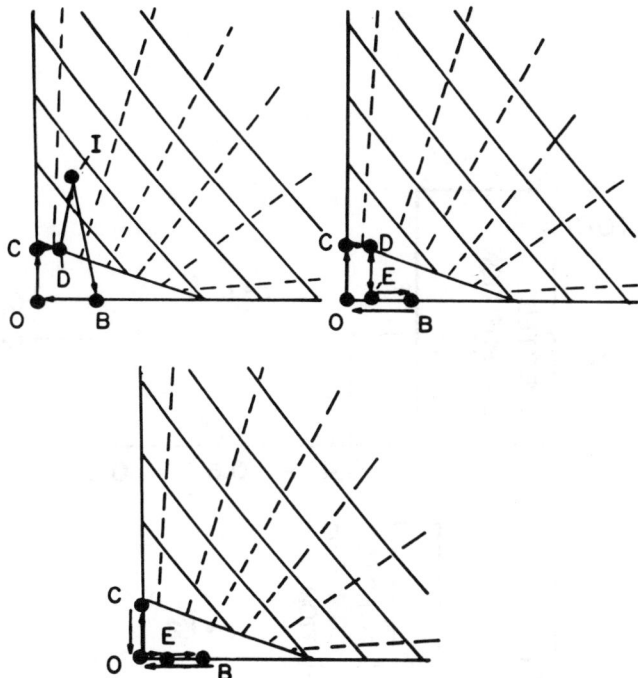

Figure 3. Chromatographic behavior of an ideal, hypothetical mixed surfactant system including mixed micelle formation. (a) Composition route development (b) Distance-time diagram (c) Column profiles. (Reproduced with permission from Ref. 8, copyright 1982, American Institute of Chemical Engineers.)

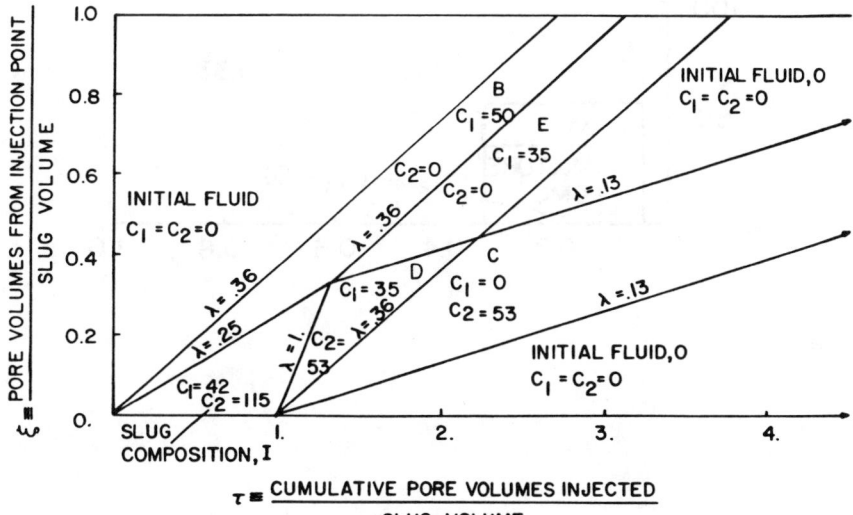

Figure 3B.

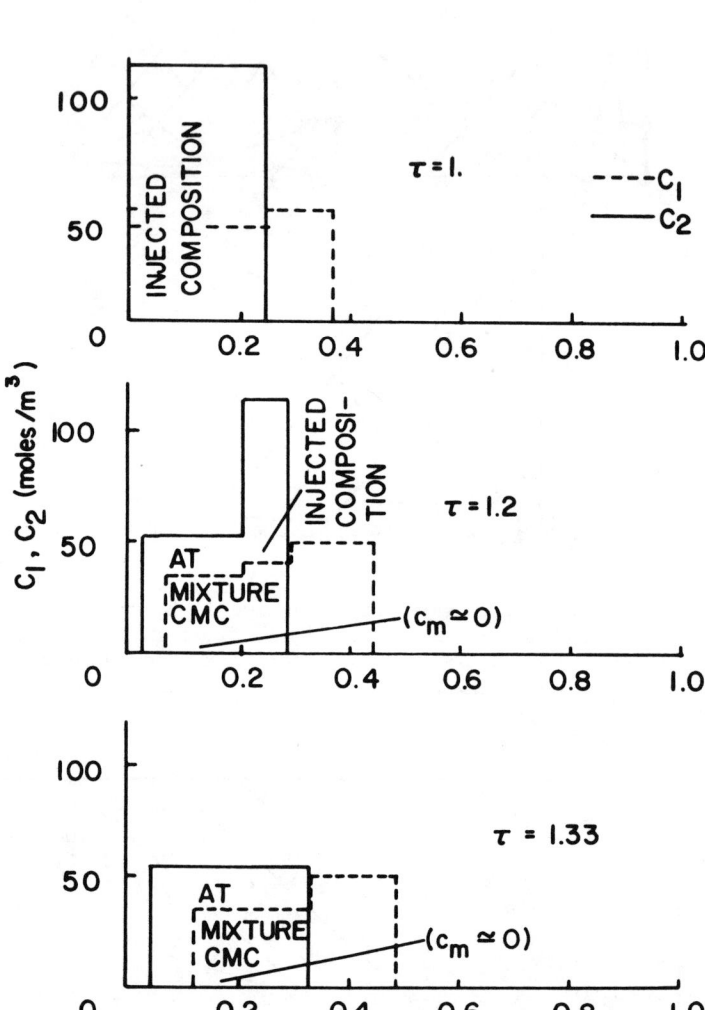

Figure 3C.

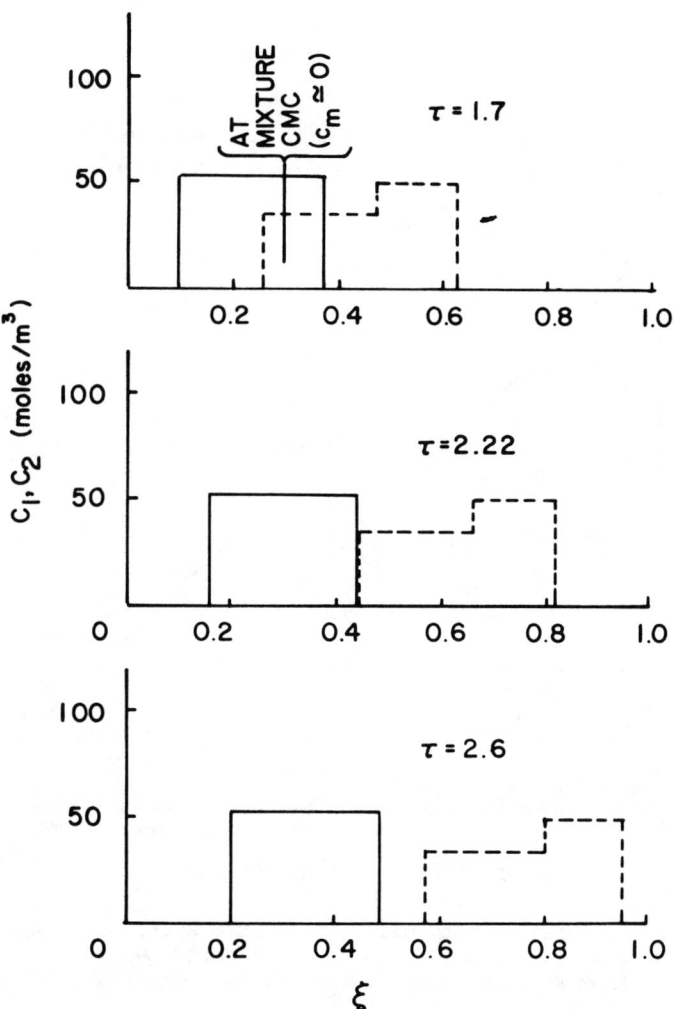

Figure 3C (Cont.).

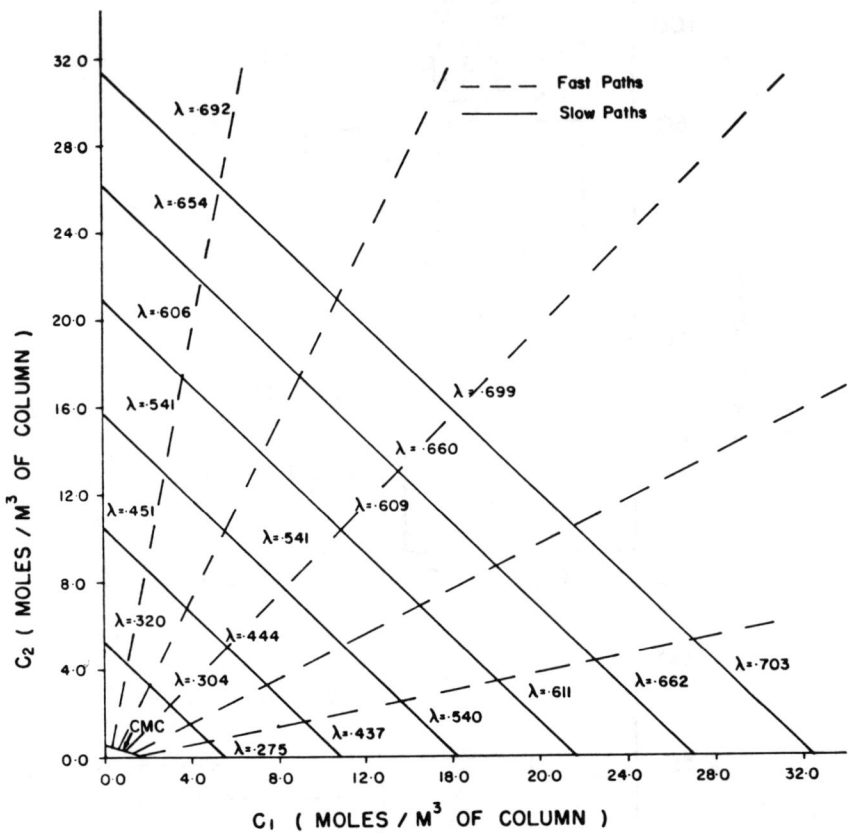

Figure 4.   Composition path grid for a mixed surfactant system using a realistic adsorption isotherm.   (Reproduced with permission from Ref. 10, copyright 1985, American Institute of Chemical Engineers.)

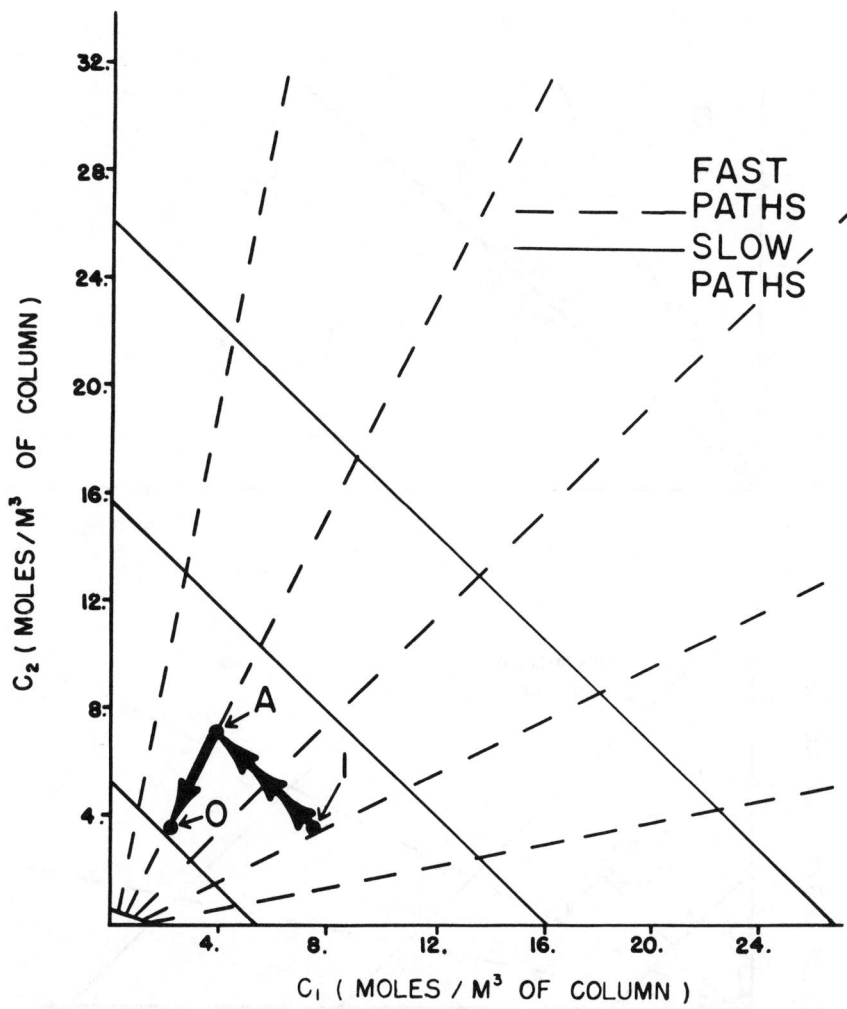

Figure 5. Chromatographic behavior of a mixed surfactant system with increase in injected composition of less strongly adsorbed component. (a) Predicted composition route development (b) Distance–time diagram (c) Observed composition route development (d) Column profiles. (Reproduced with permission from Ref. 10, copyright 1985, American Institute of Chemical Engineers.)

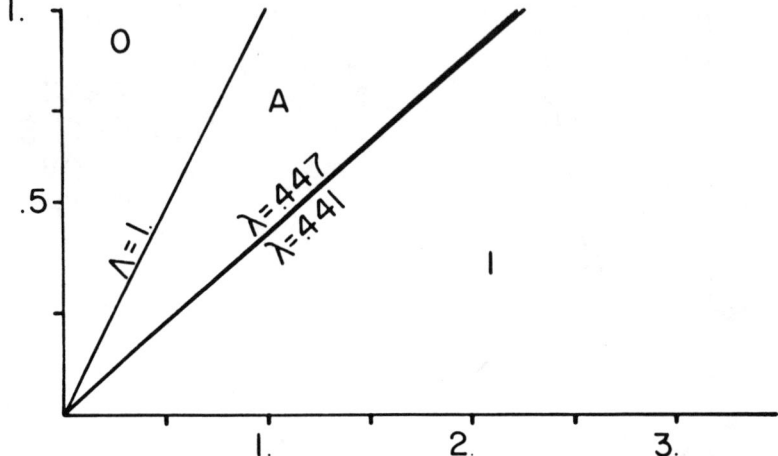

Figure 5B.

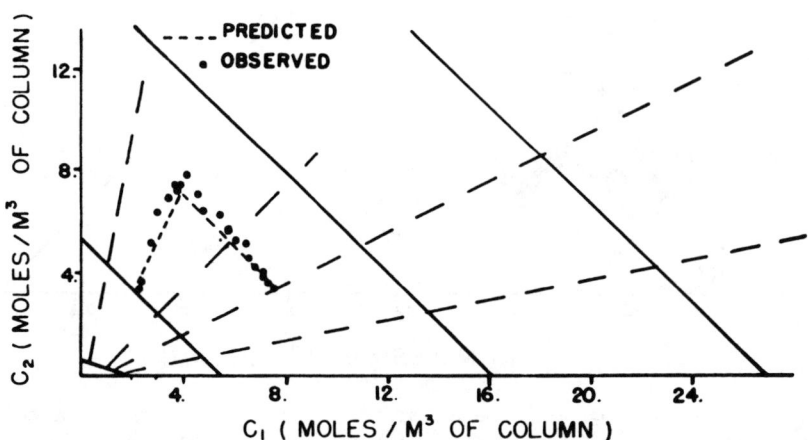

Figure 5C.

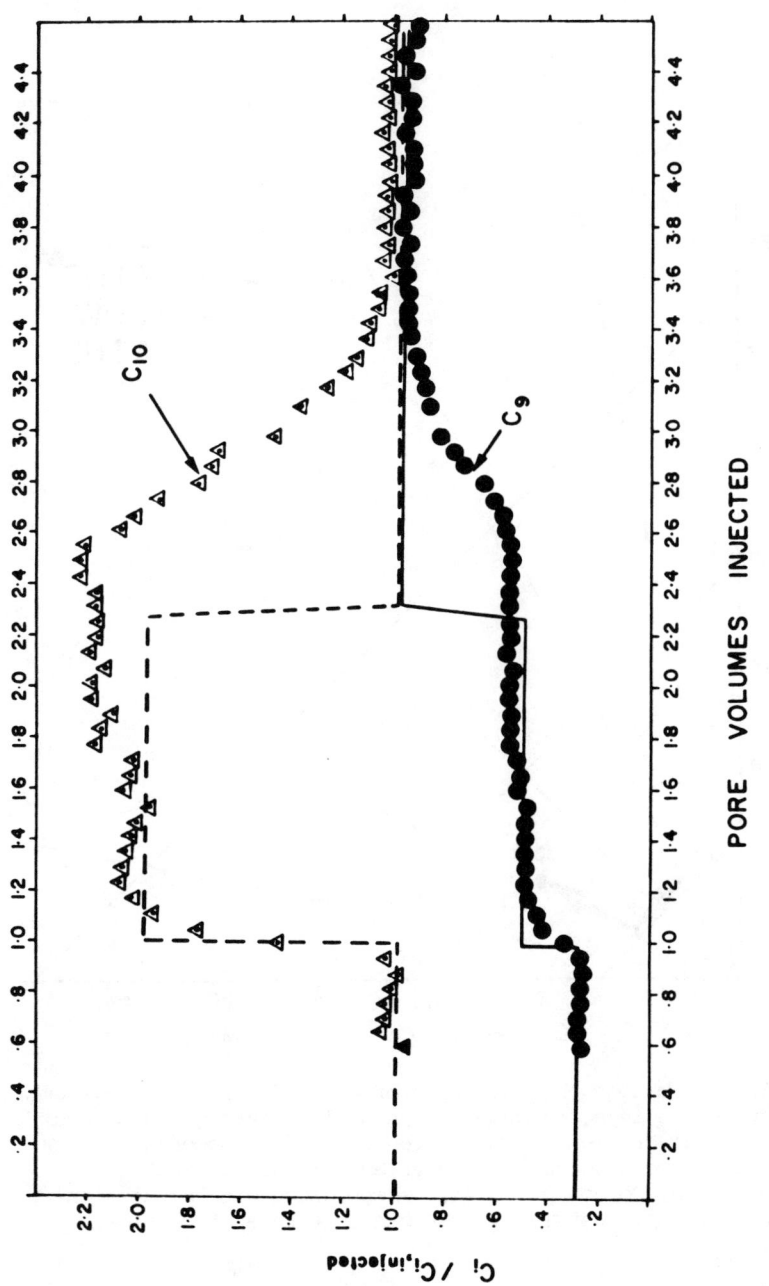

Figure 5D.

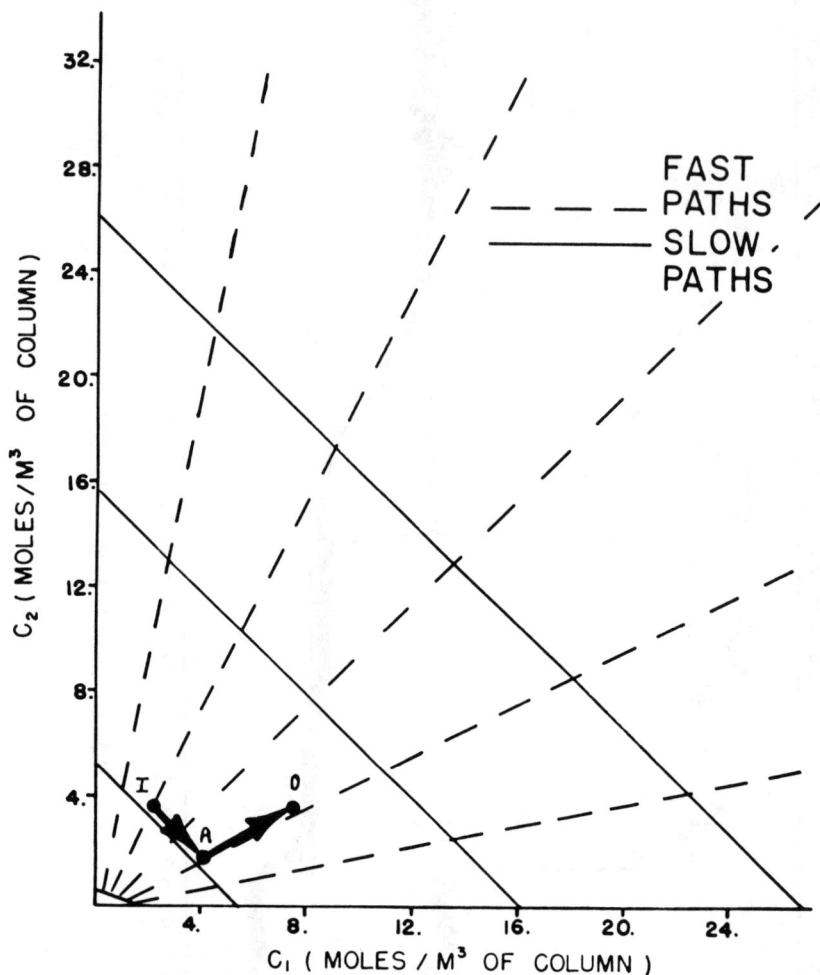

Figure 6.   Chromatographic behavior of a mixed surfactant system with decrease in injected composition of less strongly adsorbed component.   (a) Predicted composition route development (b) Distance–time diagram   (c) Observed composition route development   (d) Column profiles.   (Reproduced with permission from Ref. 10, copyright 1985, American Institute of Chemical Engineers.)

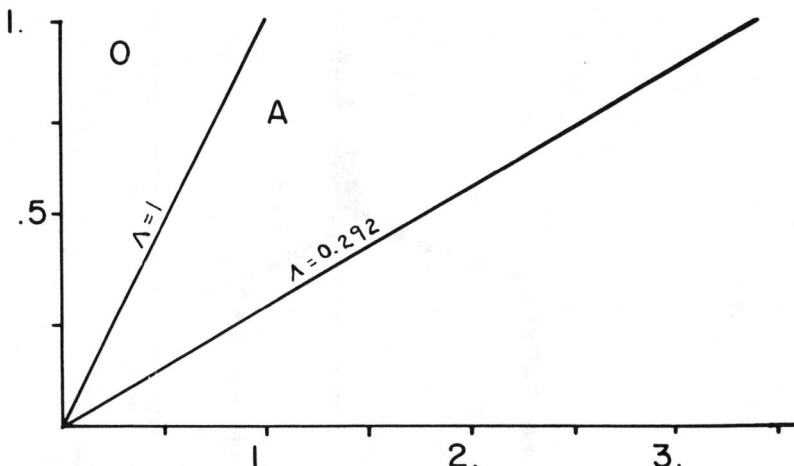

Figure 6B.

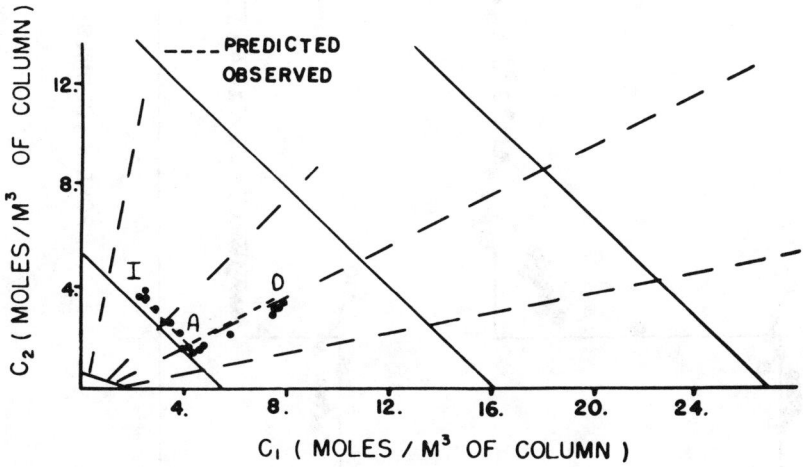

Figure 6C.

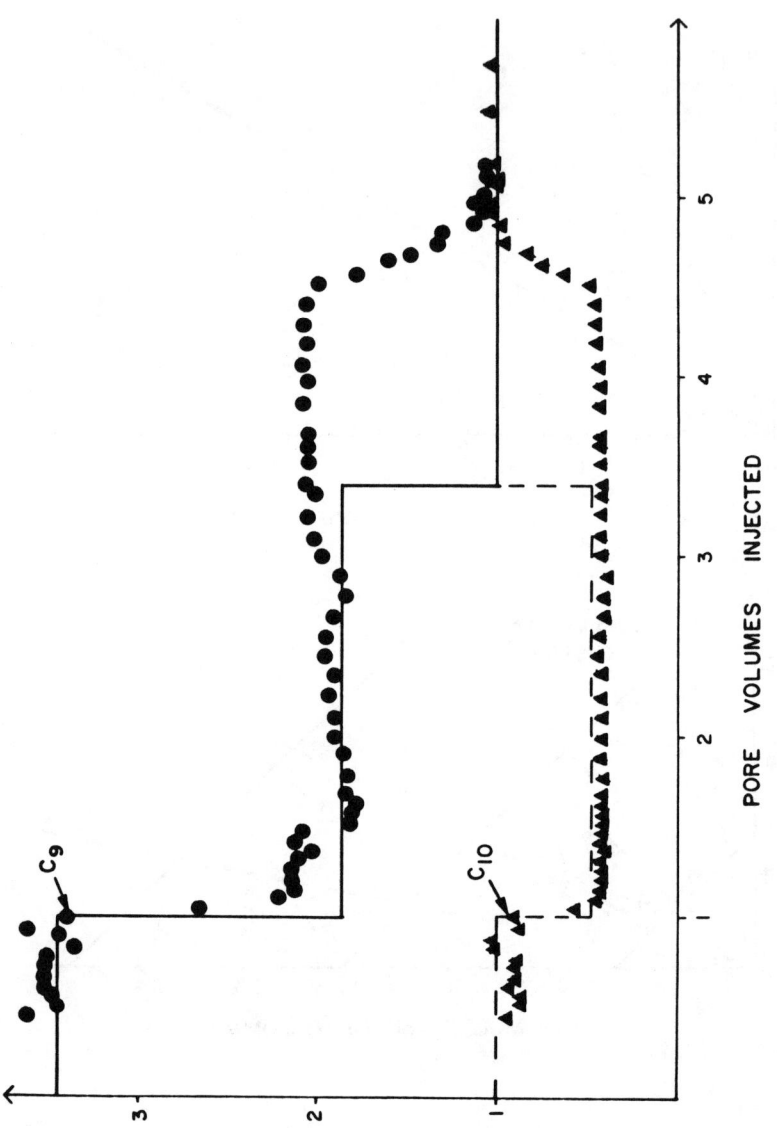

PORE VOLUMES INJECTED

Figure 6D.

Chapter 3

Technological Maturity of Sorption Processes and Sorbents

George E. Keller II

Sorption processes, including adsorption, ion exchange and chromatography, have been practiced for many years, and the first two - the subjects of this paper - have been operated commercially for many years. Uses for these processes are astonishingly broad. Technologically, however, these processes have become somewhat mature, and it can be expected that continued research on their engineering embodiments will bring less and less improvement per unit of effort expended. On the other hand, there appears to be much that can be done regarding improved sorbents, and research in this direction can be much more fruitful. For gas-adsorption technology, a new process has been conceived and developed which can greatly improve the capacity and energy efficiency of such processes. These gains are won at the expense of mechanical complexity, and it remains to be seen whether this technology is truly a harbinger of the next generation for gas adsorption.

The objectives of this paper are (i) to provide an assessment of the present state of certain segments of sorption technology and (ii) to suggest some directions in which this technology will go in the future. In this latter part the prime emphasis will be on gas adsorption. Sorption is defined here as the selective concentration of one or more components (sorbates) of either a gas or a liquid at the surface of a microporous solid or in the interior of a polymeric material. Defined in this way, sorption processes include adsorption-based processes as well as chromatographic and ion-exchange-based processes. However, since chromatographic processes are discussed and summarized in another chapter in this book, they will be mentioned only in passing here. Our concern in this paper is with both sorption processes and sorbents, with a strong emphasis on present and future industrial and other "significant-scale" applications.

The number of industrial applications for sorption processes is astonishing. As a separation technology, sorption processes are used more widely than any other non-vapor-liquid separation process for molecular separations throughout petroleum, petrochemical and chemical industries. Sorption processes also have a major presence in such diverse areas as waste treatment, water purification, pharmaceutical production, biochemical production, mineral recovery and various consumer-related applications. A representative but very truncated list of applications is given in Table I.

The background literature on sorption, sorbents and sorption processes is far too large to be covered in detail here, and instead some recent reviews and other pertinent overview articles and books are listed in the refs. 1 through 22. Especially in the area of chromatography there is a wealth of literature in various analytical-chemistry journals and books, and these should definitely be consulted for in-depth reviews.

Genesis, Progress and Maturation of Technology

In this section we will establish a basis for evaluating how mature a certain technology is. In the context of this paper, "technology" refers to a certain subset of sorption technology, such as pressure-swing adsorption (PSA), temperature-swing adsorption, liquid chromatography, etc. In addition we will make reference later to sorption agents, for example, activated carbons, molecular sieves, affinity agents, etc.

It has been noted elsewhere (23-24) that the technological progress of overall processes and products tends to follow an S curve, as shown in Figure 1. This curve shows the degree of improvement in performance of a process as a function of effort. It has also been suggested (25-26) that the S curve can represent the degree of improvement in performance of individual unit operations, and we will operate on that assumption here. Performance might be measured in terms of, for example, the cost of producing oxygen from air via PSA or the size of a chromatographic column required to purify a certain enzyme at a certain production rate. Effort might be measured in terms of research and development dollars spent.

The S curve in Figure 1 is broken into three stages. In the first stage, progress in performance is slow as a function of effort expended. This is the discovery/invention/innovation stage. During this stage the performance of the new process may even be inferior to existing, well-developed processes for accomplishing the same separation. There are some other complicating factors. First, processes in their very early stages of development are usually perceived to have significant technological risks associated with them, and hard-nosed plant managers are often reluctant to install "serial number one." In addition, in general only a few applications of the process can be envisioned. Needless to say, to get a new process past this stage takes a large amount of vision and commitment on the part of the researchers and their managers, plus a technology which truly has the potential to be a leap forward over existing technologies. The type of research needed during this stage has been called innovation research (26). It involves a heavy emphasis on invention and creative new approaches, with less concern for detailed understanding of all of the relevant phenomena. Empiricism is usually a large component of this type of research.

For those innovations which have true promise, sooner or later stage 2 - the rapid-growth stage - is reached. Here, with the basic discoveries and concepts in tow, performance begins to increase rapidly with additional increments of effort. Commercial applications burgeon as the perception grows that the process is sound both technologically and economically. Commercializations take the form not only of applications to entirely new separations but also as substitutions for and retrofittings of existing processes. The technology tends to become narrowed down to just a few physical embodiments as economic considerations cut away the unreasonable alternatives. Research during this stage tends to be a mixture of innovation and optimization. Optimization is that type of research which is aimed at making existing technologies better through a deeper and broader understanding of the various physical and chemical phenomena associated with them. Optimization research typically leans more heavily on fundamentals than does innovation research but less heavily on creativity and the search for completely new technological directions.

An example of a stage 2 process is oxygen production via PSA. The traditional technology is cryogenic distillation. Over the last 30 years, PSA, which was developed much later than cryogenic distillation, has begun to carve out a segment of that market, namely smaller plants which do not need a purified nitrogen supply at the same time. As PSA performance has improved, the oxygen production rate at which PSA can compete with cryogenic distillation has risen from just a few tons per day to perhaps in the order of 100 tons per day.

As a process continues to mature, stage 3 is finally reached. Here the technological limit or asymptote of the process is being approached. Smaller and smaller increments of improvement result for each additional increment of effort, and it becomes more and more difficult to justify additional research from an economic-payoff standpoint. Furthermore, fewer and fewer new uses for the process materialize, and the process also seems to approach a use asymptote. And even when new uses do materialize, little additional research is needed to adapt the process to these uses. Research in stage 3 tends to be almost entirely optimization-oriented. This research typically consists of fundamental studies - mathematical and experimental - and test programs on very large (and oftentimes expensive) equipment. In this stage usually one or only a very few major physical embodiments make up the vast fraction of the commercial applications. Table II summarizes the characteristics of these three stages

We have represented the three stages in Figure 1 as taking about the same amount of effort. In reality, this need not be the case. In most cases the S-curve will be somewhat skewed, leaving one stage to predominate. Obviously the best condition would be that in which stage 1 requires relatively little effort. This condition would result in rapid returns for the effort spent.

Such an analysis raises two questions with respect to sorption technology in the context of this volume. First, where do the various aspects of sorption technologies lie on the S curve? And then what are the implications for the amount and type of research which should be carried out on these various aspects? The objective of the next section is to give some preliminary answers to these questions.

Technological Maturity of Various Sorption Processes

In an earlier publication (26), the technological maturities of various separation processes were estimated by means of an opinion poll, and they are shown here in Figure 2. Again it should be pointed out that these placements are highly approximate and will be subject to some modification as more and more people's opinions are added to the database.

This figure does indicate rather clearly, though, that gas adsorption and ion exchange especially, and liquid adsorption to a somewhat lesser extent, are on their way to becoming rather mature technologies. The halcyon days of rapid improvement in these technologies are drawing to an end, according to this analysis, and we are left with the prospect of reduced rates of improvement with continuing expenditures for research.

An important question becomes that of where research should be applied to receive the most benefit from it. In the case of processes involving a mass-separating agent, research can be expended both in improving the process embodiment and in improving the mass-separating agent - the sorbent in our case. We know, however, that there can be a number of embodiments of some sorption technologies. For example, in gas adsorption there are three basic process concepts (see Figures 3 through 5). These process concepts vary primarily in the way in which desorption is effected: by temperature swing, pressure swing or by the use of a displacement agent. Presumably not all three concepts are at the same degree of technological maturity, and so the author, with some trepidation, has made some judgements regarding where these, as well as other sorption technologies, lie on the technological-maturity scale. The next three parts of this section address these maturities. Then we will discuss the possibilities for breakthroughs in sorbents.

Temperature-Swing Adsorption (TSA). TSA (see Figure 3), which can be practiced with both liquid and gas feeds, is usually only practical when the adsorbate is a small percentage (usually less than about 10 weight percent) of the stream. This is because turnaround times for such processes can range from several hours to over a day, and only when the adsorbate concentration in the feed is low can the on-stream time be a significant fraction of the total cycle time. The heat expended in desorbing the adsorbate is also usually several times the heat of vaporization of the adsorbate. Thus for carrying out bulk separations, TSA processes are not promising.
   TSA as a process embodiment is quite mature. Most TSA processes consist of rather conventional, fixed beds of adsorbent. For these, some cycle-optimization research remains to be done, but the overall economic effect of such research will be small.
   The relatively recent development of a very hard, microspherical activated carbon called bead activated carbon (BAC) has become the basis for a fluidized bed/fixed bed TSA process (27-28) which has resurrected interest in moving-bed adsorption. More recently a method of producing microspherical particles containing large amounts of zeolite and capable of being fluidized without catastrophic attrition losses (29) has extended the applicability of this process to other separations. The continued development and commercialization of moving-bed and fluidized-bed TSA processes is one of a very small number of innovative directions which TSA process embodiments can take.
   Further applications for TSA processes are predictable, even without further innovation, however. Many of these applications will lie in the areas of air-pollution and water-pollution abatement.

Pressure-Swing Adsorption (PSA). For pressure-swing adsorption (see Figure 4), the major problem is the fact that only one nearly-pure product - the less-adsorbed product usually - can be produced. The adsorbate is necessarily contaminated, unless a vacuum desorption step is used, with part of the less-adsorbed product during the purge step. (A vacuum desorption flowsheet is used to produce nitrogen from air commercially, but this process cannot produce high-quality oxygen at the same time.) Because of this two-pure-product limitation, the applications of PSA have been limited to systems for which this is not a problem. The largest uses are for air separation when only nitrogen or oxygen is required, and hydrogen upgrading in situations in which some hydrogen can be lost to the other product stream.

Process configurations and sequencing of flows between beds have been the source of much innovative development. Processes with numbers of beds ranging from one for small flows of oxygen (30) to perhaps a dozen in parallel for large-scale hydrogen purification (31) have been developed. Continued improvement in process designs have made it possible for PSA-produced oxygen to compete economically with cryogenic oxygen at production rates of up to 50 tons per day or more. Further encroachment into cryogenic oxygen's large-flow region by PSA can be expected. On the other hand, nitrogen produced by membranes will begin to encroach on PSA-produced nitrogen's domain as this newer technology improves more rapidly than does PSA. (The greater potential for improvement on the part of membrane processes is shown by its placement versus gas-adsorption processes in Figure 2.)

There is some room for improvement in cycle optimization, flow sequencing among beds and bed configuration for PSA, but the potential for improvement in economics is beginning to wane. By far the most improvement in process economics will come from the development of improved adsorbents - a point which will be discussed below.

A variant on PSA has been commercialized for removing water from azeotropes containing ethanol, for example (32). In this process a nitrogen stream is used to remove the adsorbed water by lowering its partial pressure in the vapor space. The heat for desorption is supplied by the heat released by adsorption. This heat is conveniently trapped in the bed, making the process nearly adiabatic. Further applications of this type can be expected for separations which are made difficult for distillation because of low relative volatilities.

In addition to the major markets listed above, opportunities for using PSA can be found, for example, in situations where a material must be purged from a cycle and a partially purified stream can be returned to the process. PSA becomes more and more desirable in cases in which the feed stream is available at the pressure required for separation.

Displacement-Purge Adsorption. Displacement-purge adsorption (Figure 5) solves the problem of producing two nearly pure products by using a displacement fluid which adsorbs approximately as strongly as the adsorbate to aid in partitioning the feed. Desorbing one material while adsorbing another also minimizes the temperature excursions which occur with the other cycles when used on bulk separations. Major economic problems can arise, however, from the complexity of the process and from the need to separate products from the displacement fluid. Nevertheless, in cases in which the separation is difficult by distillation, displacement-purge adsorption has found a technological home.

The Sorbex simulated-moving-bed technology developed by UOP (3) is the most outstanding example of displacement-purge adsorption for liquid-phase applications. Simulated-moving-bed technology is highly important for use in bulk separations of liquids. In such separations the adsorbent rapidly is saturated with adsorbate, and the problem becomes one of desorbing the adsorbate in the presence of unseparated feed material still in the bed. The countercurrent-flow pattern neatly circumvents this problem. This technology has been commercialized at multi-hundred-million-pounds-per-year production rates for purifying such products as normal paraffins, p-xylene, normal olefins, high-fructose corn syrup and others. The Sorbex process uses somewhat conventional fixed beds with many feed and product-drawoff points in the bed. The actual feed and drawoff points change with time to produce the simulated countercurrent contacting.

Actually ion-exchange processes can be thought of as displacement-purge processes, in which the purge stream contains the ions which replace the ions exchanged onto the resin. Uses for ion-exchange processes abound (see Table I and ref. 18). Nearly all of these involve rather conventional fixed-bed designs.

The primary vapor-phase, displacement-purge adsorption process involves the separation of normal and iso-paraffins. This process has also been close-coupled with a paraffin-isomerization process to give a product which is virtually 100 percent iso-paraffins by recycling the normal paraffin stream to extinction (33). Somewhat conventional fixed beds of adsorbents are used in this process.

Further applications for displacement-purge adsorption exist, but these will not be dependent so much on developing new process technologies as they will be on the development of new sorbents, as is discussed below.

Sorbents. If the process embodiments for the various sorption processes seem relatively constrained with respect to prospects for major improvements, the prospects for major improvements in sorbents seem to be much less so. Here are but a few examples of recent breakthroughs:

- A host of new molecular-sieve materials have been described (34-38) which offer the prospects for new selectivities for many separations.
- A host of new ion-exchange materials, based on, for example, the addition of chelating pendant groups containing oxygen, nitrogen, sulfur and phosphorus, are now available for highly selective recoveries of a wide range of metal ions (18).
- Sorbents which operate on the basis of relative rates of diffusion of different species in the pore structure of the sorbent have been commercialized (39). This concept of using diffusion-rate-based selectivity gives the researcher another degree of freedom in the quest for improved separation capabilities.
- Polymer-based sorbents have been found which swell dramatically to include many times their weight of water, and then expel almost all of this water when the temperature is increased by only a few degrees (40). Such sorbents could play key roles in concentrating biomaterials in dilute solutions.
- Ion-exchange materials are now being made in the form of membranes, woven fabrics and non-woven sheets. Inorganic materials can also be added to particles to change their density or impart magnetic properties to improve collection efficiency in fluidized-bed applications (18).
- Affinity agents which can perform separations of biomaterials are coming of age in industrial separations (41). These agents can perform hitherto unprecedented degrees of purification of proteins and other materials. The implications of using binding sites which rely both on specific chemical interactions with the sorbate and on the geometry of the adsorbate are major for many biomaterials as well as for many separations of stereoisomers.

Many more examples could be given, but these are perhaps enough to make the point that the possibilities for new sorption processes based on new sorbents truly are impressive - much more impressive than the opportunities for new process embodiments for these sorption processes. The degrees of freedom for selection of sorbents are simply much larger in number than the degrees of freedom left for improvement of process flowsheets, equipment details and operating strategies. Thus the main areas for fruitful research for new processes and applications are in materials science, including its inorganic, organic (polymer) and biological aspects. Improvement of process embodiments are likely to show less promise for major improvements in economics.

Limiting Conditions for Gas-Adsorption Technology

In spite of the fact that we have just concluded that sorbent research is more fruitful than process-embodiment research, there are nevertheless possibilities for significant improvements in this latter area also. In this section we return to gas-adsorption technology and ask the question as to whether we can begin to see an "asymptotic" technology - a process embodiment which would be the best imaginable - for it. From the last section we also learned that all of the three basic gas-adsorption processes have shortcomings which keep them from major growth in use, and a breakthrough in process concept will be necessary to produce this growth. The process challenge is to develop a relatively simple device which, without the use of a carrier gas or displacement medium, would do the following:

- Effect high degrees of separation between key components of the feed,
- Increase the volume of gas separated per unit time per unit of adsorbent by at least one order of magnitude over those for PSA, and
- Accomplish these goals with energy inputs per unit of feed no higher than those for PSA and preferably lower.

A new process was described recently (<u>42</u>) which seems to accomplish all of these goals simultaneously. This process is shown in Figure 6. Pistons in cylinders at each end of the bed are used to drive gas back and forth across the bed at very high rates. The back-and-forth motion of the gas is also synchronized with a cycling pressure to produce the separation, so that strongly-adsorbed product is produced at one end of the bed and weakly-adsorbed product is produced at the other end. The process uses a criterion that, when the gas flow is in one direction, the mass flux is into the adsorbent particles, and when the gas flow is in the opposite direction, the mass flux is out of the adsorbent particles. Rates of cycling are high enough (in the range of several seconds to less than one second per cycle) that there is substantial diffusion resistance in the adsorbent, so that the flux wave does not follow the pressure wave, i.e., even after the pressure wave reaches its maximum and begins to decline, gas is still entering the adsorbent. This is shown schematically in Figure 7. To create the desired flow/flux phenomenon requires the use of two pistons of different displacements and operating out of phase. The small-displacement (or small-stroke) piston "leads" the large-displacement piston, i.e., the small-displacement piston begins its movement toward the bed while the other piston is still moving outward, and then the small-displacement piston begins to move outward while the other is still moving inward. The effect of the phase angle between the pistons is shown in Figure 8. Here it appears that the optimal phase angle is about 30 to 45 degrees. Table III also shows that the process can produce two quite pure products. (Actually the balance of the oxygen stream is nearly all argon, which concentrates with the oxygen.) The productivities at which the products are produced are considerably higher than values for more conventional processes. It is also of interest that the pistons allow for recovery of part of the energy expended to compress the gas, since as the pistons retreat, the gas in the bed returns work to the motor driving the pistons.

This adsorption process would seem to fulfill the criteria listed above, but at a cost. The cost is generated in increased process complexity, which results in the potential for increased investment. It remains to be seen whether such technology can begin to supplant the mechanically simpler processes we operate today.

It seems almost inevitable that as separation technologies are driven to their limits, their embodiments will become more and more mechanically complex. This is primarily because of the need to increase the rates of mass transfer, heat transfer and phase disengagement. For example, in distillation and solvent extraction, centrifugal forces have been used to produce high-gravitational fields which both increase mass-transfer rates and also facilitate phase separation (43-45). So far, however, these technologies are used only sparingly commercially and have certainly not displaced the much simpler column technologies which have been used for decades. Thus, whether or not "mechanically driven" solutions for approaching technological asymptotes for gas adsorption as well as for other separation processes is still an open question, and another decade or two will be required to answer the question definitively.

Literature Cited

1. Anderson, R. E.   In Handbook of Separation Techniques for Chemical Engineers; Schweitzer, P. A., Ed.; McGraw-Hill Book Co.: New York, 1979; p 1-359.
2. Ausikaitis, J. P.; Myers, A. L., Eds. Adsorption and Ion Exchange: Recent Developments; AIChE Symposium Series No. S-242, American Institute of Chemical Engineers: New York, 1985.
3. Broughton, D. B.   In Kirk-Othmer Encyclopedia of Chemical Technology; Wiley-Interscience: New York, 1978; 3rd Edition, Vol. 1; p 563.
4. Chi, C. W.; Cummings, W. P.   In Kirk-Othmer Encyclopedia of Chemical Technology; Wiley-Interscience: New York, 1978; 3rd Edition, Vol. 1; p 544.
5. Clearfield, A., Ed. Inorganic Ion Exchange Materials; CRC Press: Boca Raton, FL, 1982.

6. Faust, S. D.; Aly, O. M. <u>Adsorption Processes for Water Treatment</u>; Butterworth Publishers: Stoneham, MA.
7. Flank, W. H., Ed. <u>Adsorption and Ion Exchange with Synthetic Zeolites</u>; ACS Symposium Series No. 135, American Chemical Society: Washington, 1980.
8. Hutchins, R. A. In <u>Handbook of Separation Techniques for Chemical Engineers</u>; Schweitzer, P. A., Ed.; McGraw-Hill: New York, 1979; p 1-415.
9. Keller, G. E.; Anderson, R. A.; Yon, C. M. " In <u>Handbook of Separation Process Technology</u>; Rousseau, R. W., Ed.; John Wiley & Sons: New York, 1987; p 644.
10. Kovach, J. L. In <u>Handbook of Separation Techniques for Chemical Engineers</u>; Schweitzer, P. A., Ed.; McGraw-Hill Book Co.: New York, 1979; p 3-3.
11. Ma, Y. H., Ed. <u>Recent Advances in Adsorption and Ion Exchange</u>; AIChE Symposium Series No. S-219; American Institute of Chemical Engineers: New York, 1982.
12. Ma, Y. H.; Cooney, D. O.; Hines, A. L., Eds. <u>Adsorption and Ion Exchange-'83</u>; AIChE Symposium Series No. S-230; American Institute of Chemical Engineers: New York, 1983.
13. Myers, A. L.; Belfort, G., Eds. <u>Fundamentals of Adsorption</u>; AIChE Symposium Series No. P-39; American Institute of Chemical Engineers: New York, 1984.
14. Naden, D.; Street, M., Eds. <u>Ion Exchange Technology</u>; Ellis Horwood Ltd.: Chichester, UK, 1986.
15. Perrich, J. R., Ed. <u>Activated Carbon Adsorption for Wastewater Treatment</u>; CRC Press: Boca Raton, FL, 1981.
16. Ruthven, D. M. <u>Principles of Adsorption and Adsorption Processes</u>; John Wiley & Sons: New York, 1984.
17. Sherman, J. D.; Vermeulen, T., Eds. <u>Adsorption and Ion Exchange - Progress and Future Prospects</u>; AIChE Symposium Series No. S-233; American Institute of Chemical Engineers: New York, 1984.
18. Streat, M.; Cloete, F. L. D. In <u>Handbook of Separation Process Technology</u>; Rousseau, R. W., Ed.; John Wiley & Sons: New York, 1987; p 697.
19. Vermeulen, T. In <u>Kirk-Othmer Encyclopedia of Chemical Technology</u>; Wiley-Interscience: New York, 1978; 3rd Edition, Vol. 1, p 531.
20. Vermeulen, T.; Klein, G.; Hiester, N. K. In <u>Chemical Engineers' Handbook</u>; Perry, R. H.; Chilton, C. H., Eds.; McGraw-Hill: New York, 1973; 5th edition, p 16-1.
21. Wankat, P. C. <u>Large-Scale Adsorption and Chromatography</u>; CRC Press: Boca Raton, FL, 1986; 2 Volumes.
22. Yang, R. T. <u>Gas Separation by Adsorption Processes</u>; Butter-worth Publishers: Stoneham, MA, 1987.
23. Foster, R. N. <u>Chemtech</u> December 11, 1983, 720.
24. Foster, R. N. <u>Innovation: The Attacker's Advantage</u>; Summit Books: New York, 1986.

25. Foster, R. N.; Linden, L. H.; Whiteley, R. L.; Dantrow, A. M.   <u>Improving the Return on Research and Development</u>; Industrial Research Institute: 100 Park Avenue, New York, 1984.
26. Keller, G. E.   <u>Separations: New Directions for an Old Field</u>; AIChE Monograph Series; American Institute of Chemical Engineers: New York, 1987; Vol. 83.
27. Sakaguchi, Y.   <u>Chemical Economy and Engineering Review</u> 1976, <u>8</u> (12), 36.
28. <u>PURASIV HR for Hydrocarbon Recovery</u>; Union Carbide Corporation: Danbury, CT.
29. Acharya, A.; BeVier, W. E. U. S. Patent 4 526 877, 1985.
30. Keller, G. E.; Jones, R. L.   In <u>Adsorption and Ion Exchange with Synthetic Zeolites</u>; Flank, W. H., Ed.; ACS Symposium Series No. 135; American Chemical Society: Washington, 1980; p 275.
31. Cassidy, R. T.   In <u>Adsorption and Ion Exchange with Synthetic Zeolites</u>; ACS Symposium Series No. 135; American Chemical Society: Washington, 1980; p 247.
32. Garg, D. R.; Ausikaitis, J. P.   <u>Chem. Eng. Prog.</u> 1983, <u>79</u> (4), 60.
33. Symoniak, M. F. <u>Hydrocarbon Processing</u> May, 1980, 110.
34. Kokotailo, G. T.; Meier, W. M.   In <u>Properties and Applications of Zeolites</u>; Townsend, R. P., Ed.; Chemical Society: London, 1979; Special Publication No. 33.
35. Flanigen, E. M.; Bennett, J. M.; Grose, R. W.; Cohen, J. P.; Patton, R. L.; Kerchner, R. M.; Smith, J. V. <u>Nature</u> 1978, <u>271</u>, 512.
36. Wilson, S. T.; Lok, B. M.; Messina, C. A.; Cannan, T. R; Flanigen, E. M.   <u>J. Am Chem. Soc.</u> 1982, <u>104</u>, 1146.
37. Messina, C. A.; Lok, B. M.; Flanigen, E. M.   U. S. Patent 4 554 143, 1985.
38. Wilson, S. T.; Flanigen, E. M.   U. S. Patent 4 567 029, 1986.
39. Knoblauch, K.   <u>Chemical Engineering</u> November 6, 1978, 87.
40. Cussler, E. L.; Stokar, H. R.; Varberg, J. E.   <u>AIChE Journal</u> 1984, <u>30</u> (4), 578.
41. Janson, J.-C.   <u>Trends in Biotechnology</u> 1984, <u>2</u> (2), 31.
42. Keller, G. E.; Kuo, C. H. A.   U. S. Patent 4 354 859, 1982.
43. Ramshaw, C.; Mullinson, R. H.   U. S. Patent 4 283 255; 1981.
44. Ramshaw, C.   <u>The Chem. Engr.</u> February, 1983, 13.
45. Barson, N.; Beyer, G. H.   <u>Chem. Eng. Prog.</u> 1953, <u>49</u> (5), 243.

Table I.   Examples of Commercial Adsorption and
Ion Exchange Separation Applications

Adsorption: Liquids	Ion Exchange
Normal/iso-paraffins, aromatics	Removal of various ions
Normal olefins/normal paraffins	from drinking water
Fructose/glucose	and industrial water
p-Xylene/o-xylene, m-xylene	Purification of various
Drying of organics	raw sugar solutions
Waste water cleanup	and other food-
Drinking-water purification	industry streams
Decolorizing of petroleum	Purification of bio-
fractions, syrups, etc.	products such as
Adsorption: Gases	vitamins, amino acids,
Normal/iso-paraffins, aromatics	proteins and enzymes
Oxygen, nitrogen from air	Hydrometallurgical
Hydrogen recovery	recovery of actinides,
Organics from vent streams	noble metals, transi-
Drying of gas streams	tion metals, etc.
Water removal from	Purification of organic
azeotropes	solvents

Table II.   Description of Stages in Maturation
of Technology

Stage	Character of Research	Technologically, What is Going On	Use-Wise, What is Going On
1	Innovation	Discovery. Technology oftentimes not as good economically as established alternatives. Many possible technological directions. Requires technology advocates.	One or only a few uses envisioned. Reluctance on the part of plants to commercialize.
2	Innovation/ optimization	Technological directions become focused. Rapid improvement in performance.	Rapid expansion in number of uses commercialized. Acceptance by plants. Becomes part of "conventional wisdom."
3	Optimization	Technology becomes standardized to one or a very few embodiments. Fill-in-the-gaps research. Slowdown in the rate of improvement.	Many large-scale uses but sharp decline in the rate of applications to new uses.

Table III.   Comparison of Product Purities and
Productivities

| Product Purity, mole % | | Productivity, kg/kg adsorbent-day | |
N_2 (Long-Stroke Piston)	O_2 (Short-Stroke Piston)	N_2	O_2
89.0	70.2	10.4	1.8
92.0	82.0	8.4	1.8
95.7	89.2	7.2	1.6
97.8	94.0	5.2	1.5
99.9	95.0	4.4	1.4

Conditions:
  Pistons: 4.5 cm bore, stroke lengths of 7.6 and 2.5 cm
  Phase angle: 45° (short-stroke piston leads)
  Adsorbent: 83 g of 13X molecular sieve, 40x80 mesh
  Feed pressure: 200 KPa
  Cycle frequency: 30 rpm

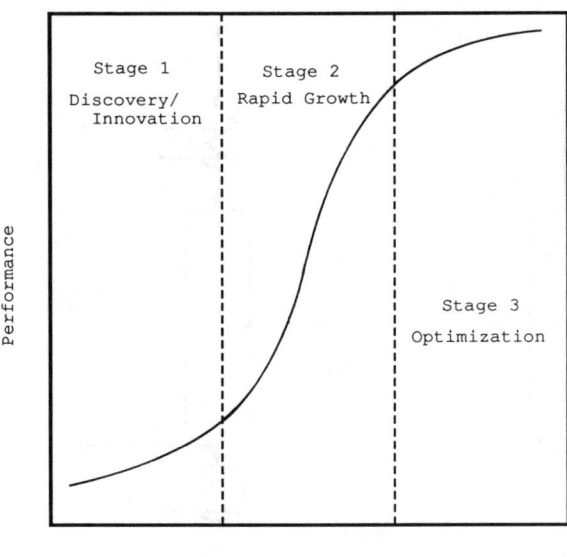

Figure 1.   Maturation of Technology with Effort.

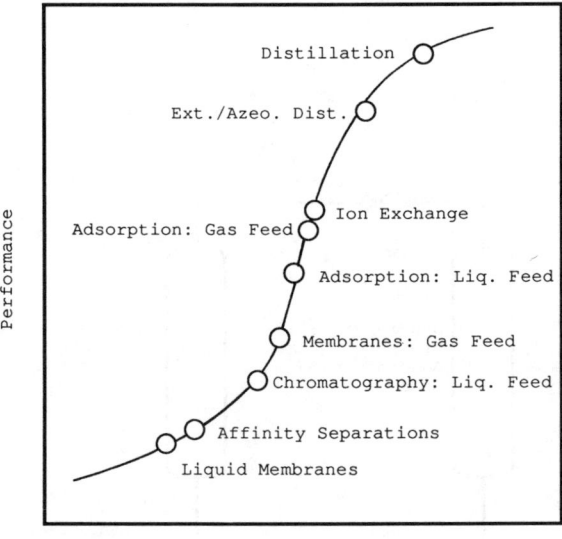

Figure 2.   Technological Maturities of Various Separation Processes.

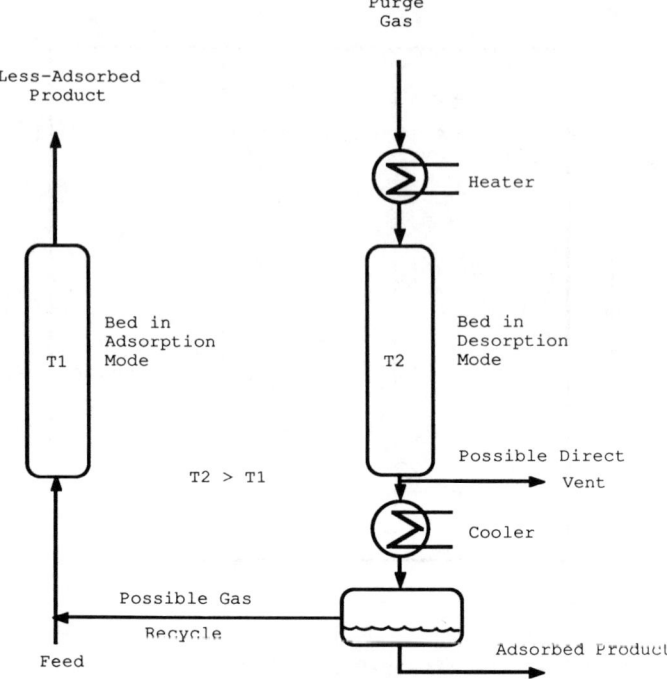

Figure 3.   Schematic of Temperature-Swing-Adsorption (TSA) Process.

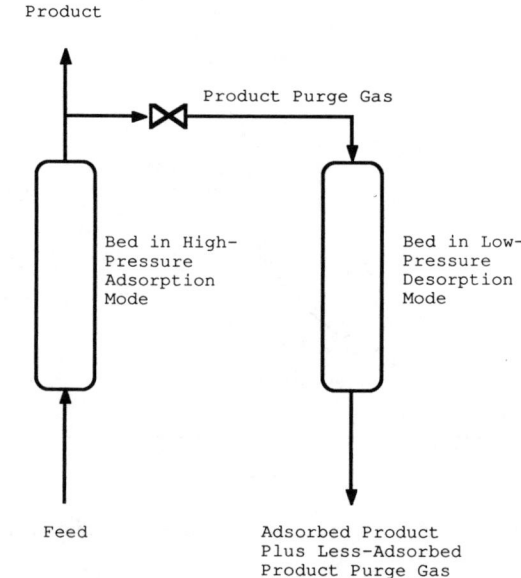

Figure 4.   Schematic of Pressure-Swing-Adsorption (PSA) Process.

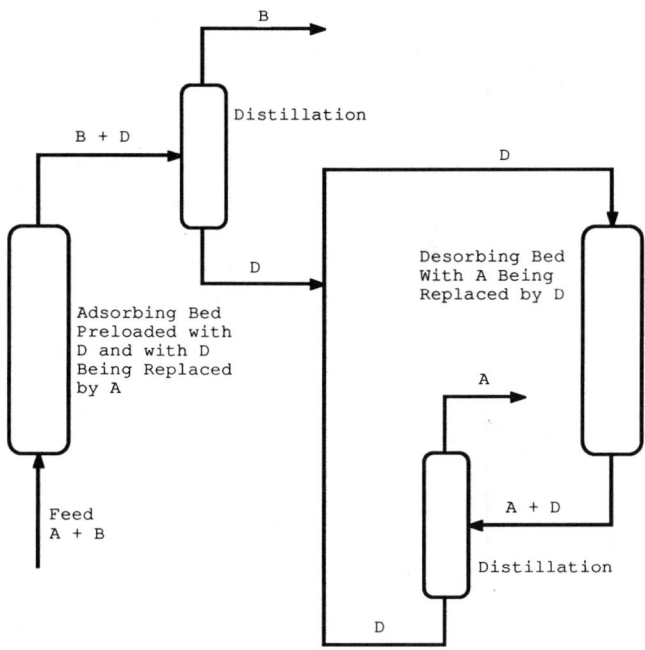

Figure 5.   Schematic of Displacement-Purge-Adsorption Process.

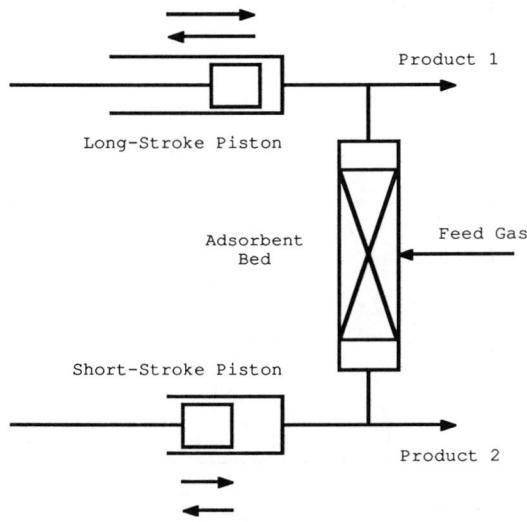

Figure 6.   Schematic of Piston-Driven PSA Process.

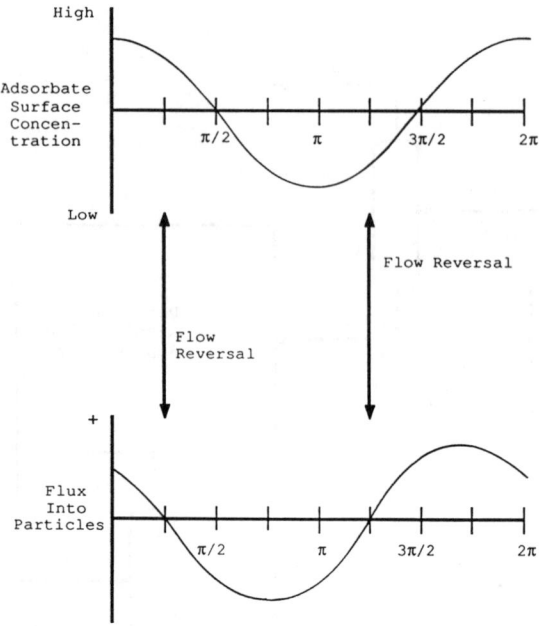

Figure 7.   Comparison of Surface-Concentration and Flux Waves.

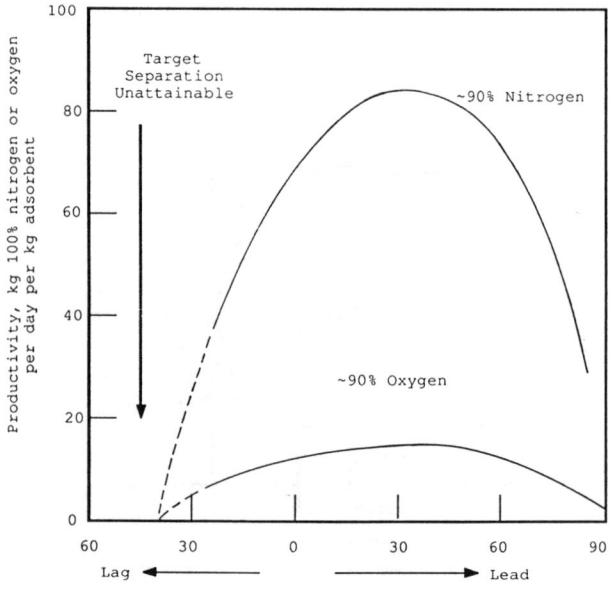

Phase Angle, short vs. long piston

Figure 8.   Effect of Piston Phase Angle on Productivity During
Air Separation.

Chapter 4

Kinetic Separation of Air by Pressure Swing Adsorption

Z.J. Pan
R.T. Yang
J.A. Ritter

The kinetic separation of air for the production of
nitrogen (and argon) by pressure swing adsorption
was studied using 4A zeolite sorbent. The basic
underlying principles of kinetic separation were
studied by examining various PSA cycles. Nearly
equal lengths of time should be alloted in the
cycle for the fast-diffusing component (oxygen) to
adsorb and desorb, while this time should be
sufficiently short to prevent the adsorption of
substantial amounts of the slow-diffusion species
(nitrogen and argon). The most efficient
separation was given by the cycle consisting of
the following steps: feed pressurization (I),
cocurrent bloedown (II), countercurrent blowdown (III)
and delay (IV). Steps I and II allowed oxygen
adsorption while Step II yielded the desired product.
Step III desorbed oxygen. Nitrogen was the gas
desorbed during delay and more importantly, was kept
in the column which was the product in the ensuing
cycle. Thus the delay step increased the product
recovery. Cocurrent blowdown also enhanced product
recovery. The functions and mechanisms of delay and
cocurrent blowdown are both different from those
employed in equilibrium separations.

Pressure swing adsorption (PSA) is a major development in gas adsorption technology (1). In PSA the sorbent is periodically regener-
ated by reducing the total pressure. The cycle time is short,
usually in minutes or even seconds, due to the possibility of rapid
pressure reduction. Consequently large throughputs can be achieved
and an increasing number of new separations using PSA are being
commercialized.
    The adsorptive separation is achieved by one of the three
mechanisms: sterid, kinetic or equilibrium effect. The steric
effect derived from the molecular sieving property of zeolites. In

this case only small and properly shaped molecules can diffuse into the adsorbent, whereas other molecules are totally excluded. Kinetic separation is achieved by virtue of the differences in diffusion rates of different molecules into the sorbent. By far, however, most processes operate through the equilibrium (or competitive) adsorption of the mixture, and hence are called equilibrium separation processes.

Commercial separation processes using molecular sieve zeolites are mainly equilibrium processes. They are only two major steric separation processes: drying with 3A zeolite and the separation of normal paraffins from iso-paraffins and cyclic hydrocarbons using 5A zeolite.

The only kinetic separation process that is commercialized is air separation (mainly for producing nitrogen) using molecular sieve carbon. Molecular sieve zeolites, despite their enormous potential, have yet to be used commercially for kinetic separation.

The possibility of kinetic separation of air by 4A zeolite was first realized by Keller at Union Carbide about a decade ago (2) and was briefly mentioned in their patent (3). The diffusion time in 4A zeolite crystals is in seconds for O_2 and minutes for N_2. Thus N_2 can be generated by PSA with cycle times lying between the two diffusion times. Such rapid cycles can yield a high productivity (throughput per unit amount of sorbent) and are ideally suited for aircraft inerting purposes, where the required N_2(Ar) concentration is approximately 92%. Following the lead of the work at Union Carbide, there have been extensive research activities for developing inerting units by kinetic separation (PSA cycles using molecular sieve carbon have longer cycle times and give lower productivities) (4,5,6).

In this paper we report new kinetic separation cycles and provide a basic understanding of the cyclic process.

Experimental

A highly automated single-column apparatus, similar to that described elsewhere (1), was used in this study. The column was 2.05 cm inside radius by 60.6 cm long, containing 520 g 4A zeolite pellets (after regeneration). The PSA cycles were operated by solenoid valves which, in turn, were operated by timers controlled by a micro-processor. The effluent flowrates were controlled by fine needle valves. A product surge tank was used in experiments where product pressurization was required. Pressure and temperature of the column were continuously recorded. The effluent compositions were analyzed by both gas chromatography and a paramagnetic oxygen analyzer.

The sorbent was 4A zeolite pellets, 1/16 inch, from Molecular Sieves Department of Union Carbide Corporation. Laboratory compressed air was passed through beds of activated carbon and drierite to remove oil vapors and moisture. The column was desorbed at 350°C in helium flow for 3 hrs. prior to the series of PSA experiments. The Amounts of cumulative impurities in the column (moisture, CO_2, etc.) were judged to be negligible since the PSA results could be reproduced upon the end of the experimental series. The bed was evacuated and purged with He between experiments.

A cyclic steady state was reached after about 35 cycles. All results given in this paper were taken at cyclic steady states.

Because the amounts adsorbed and desorbed during each cycle were small, the cyclic temperature swings were within 2-3°C, which could be considered as isothermal.

Diffusivities and Isotherms

The equilibrium isotherms for N_2 and O_2 on 4A zeolite at 25°C and are shown in Figure 1. The Ar isotherm is not shown but is near that of O_2. As for most gas-solid systems, the thermodynamic equilibria are not favorable for kinetic separation because the stronger adsorptive (N_2) has a lower diffusivity.

The diffusivities of O_2, N_2 and Ar in 4A zeolite crystals have been measured by a number of groups using different techniques. These data are given in Table I. From these data, the characteristic diffusion time (r^2/D where r is crystal radius and D is diffusivity) for O_2 is 10-100 x, and that for N_2 and Ar is about 2-3 orders of magnitude longer. The characteristic diffusion time for O_2 in crystals is also much shorter than that of O_2 in the pellets. Thus the diffusion of O_2 in crystals is the dominant step in the kinetic separation cycle.

Cycle Strategy and Cycles Studies

Consider the separation of a binary O_2/N_2 mixture, O_2 being the fast-diffusing component. The following principles should be obeyed for kinetic separation:

(1) A sufficiently long time must be provided during each cycle for O_2 to adsorb; while this time must be short enough to prevent the adsorption of a large amount of N_2.

(2) A sufficiently long time must be provided during each cycle for O_2 desorption.

(3) Since some N_2 is also adsorbed during the O_2 adsorption step, a cycle step should also be allowed for N_2 desorption. Ideally, O_2 and N_2 desorption should be accomplsihed in two separate steps. The N_2 desorption step is important for a high N_2 recovery.

Following these principles, all separations that can be accomplished by equilibrium separation can also be accomplished by kinetic separation using a proper molecular sieve sorbent. However, equilibrium separation is suited for the production of the weak adsorptive in the mixture while kinetic separation is suited for the strong adsorptive (which is usually the slow-diffusing component).

PSA cycle strategies were developed based on the preceding principles. As in all PSA cycles, adsorption is accomplished during the steps of pressurization and high-pressure feed, and desorption is accomplished by the blowdown steps and purge-evacuation steps. The purge and evacuation steps will not be considered here because purge would consume the desired product (N_2) and substantially lower the product recovery, and the benefits of using evacuation would not likely justify the effort.

A step-by-step development of the kinetic separation cycle will be given. The first strategy is to allot equal time for adsorption and desorption in the cycle. The direction of blowdown is then considered. As the results will show, the blowdown direction is crucially important in kinetic separation, and more important, O_2 (the fast-diffusing component) is desorbed in cocurrent blowdown (which is cocurrent to the feed direction) and N_2 is desorbed in countercurrent blowdown. Furthermore, advantages will be shown for using a delay step, which not only effectively desorbs N_2, but also increases the N_2 product recovery by keeping the desorbed N_2 in the bed.

Following the preceding strategy, the cycles studied are shown in Figure 2.

Results and Discussion

The separation results are given by three inter-related sets of data: product purity, product recovery and product throughput. As mentioned, the slow-diffusing component (N_2) is the desired product. (Equilibrium separation should be used when O_2 is the desired product.) Product recovery is defined as the fraction of the component in the feed that is recovered in the product stream. What follows is a step-by-step development of the kinetic separation cycle and an illustration of the function that each cyclic step performs.

Selection of Cycle Time. The first parameter to be determined for the kinetic separation cycle is the proper length of gas residence time to allow only the fast-diffusing component (O_2) to adsorb. The question is reduced to the proper cycle time once the feed rate is fixed. The proper cycle time was determined using the simplest cycle, Cycle A shown in Figure 2, which allows approximately equal times for adsorption and desorption.

To determine the proper residence time, 14 PSA cycles with various combinations of feed flowrates and total cycle times were studied. The results are summarized in Table II. The total feed amount per cycle included two portions: feed used in pressurization (Step I) and that used in feed (Step II). The N_2 + Ar product was given by Step II effluent. The product rate was controlled by adjusting a needle valve in the effluent line; the feed rate in Step II was not controlled or measured. The cyclic pressure histories for all 14 runs were the same as shown in Cycle A (Figure 2) except their time scales were different. The total cycle time varied from 20 to 360 s. A half of the cycle time was spent for Steps I and II when adsorption took place, while the other half was intended for desorption.

To evaluate the separation, comparisons should be made for three results: N_2 + Ar product (throughput) rate, N_2 + Ar product purity, and N_2 + Ar product recovery. To simplify the evaluation, product purity was plotted against product recovery from the data of all 14 runs as shown in Figure 3. These curves may be considered as performance curves. Better performance is given by higher curves is the product rates are similar.

Based on the performance curves given in Figure 3, the cycle time of 40 s appeared to be optimum for the given kinetic separation. The separation deteriorated at longer cycle times. The performance of cycles with 20 s cycle time appeared to be inferior than that of 40 s cycles. However, the product rate increased with shorter

cycles. Thus, if product rate is the dominant factor for consideration, shorter cycles would be advantageous.

A wealth of information can be gained from the data given in
Table II. An apparent but oversimplified conclusion is that 20 s
time for both O_2 adsorption and desorption, as in the 40 s cycles,
appeared to be the optimum, and longer cycle times would result in
N_2 adsorption. However, the bed dynamics were far more complicated
than this simple picture. The residual gas phase in the void spaces
at the end of each cycle (after blowdown to 0 psig) was approximately
0.6 ℓ STP. Therefore, the actual residence time of the gas that was
contained in the N_2 + Ar product was longer than the total cycle time
for all runs. Moreover, at the end of the high-pressure feed step,
the gas near the discharge end was enriched in N_2 + Ar. During the
ensuing countercurrent blowdown step, this gas reversed its direction
and purged toward the feed end. This self-purge action served the
same function as the purge step in the Skarstrom cycle (1) without,
however, consuming the product for purge as required in the Skarstrom
cycle.

Improvement by Cocurrent Blowdown. In Cycle A, countercurrent blowdown immediately followed the high-pressure feed step. Thus the
mxiture near the feed end of the column was immediately discharged
without time for adsorption. This portion of the high-pressure
feed was consequently wasted. A remedy for this is to insert a
cocurrent blowdown step before countercurrent blowdown, thus allowing
time for O_2 to diffuse. However, the effluent from this cocurrent
blowdown step was not as concentrated in N_2 + Ar as that from the
high-pressure feed step. Subsequently the high-pressure feed step
was eliminated in order to increase the product purity at the
expense of feed throughput. Such a cycle is shown as Cycle B in
Figure 2.

The results of 7 runs using Cycle B are given in Table III. The
feed amount in each cycle was exclusively incurred in the feed
pressurization step. The amount of the gas phase contained in the
voids of the bed at the end of pressurization (at 25 psig) was
appproximately 1.7 ℓ STP. The difference between this amount and
the feed amount was the amount that was adsorbed during the time of
the pressurization step. It is interesting to note the large increase in feed amount as the cycle time was increased (Table III,
see Feed Amount). The amount adsorbed during pressurization seemed
to increase linearly with pressurization time as it increased from 10
s (40 s cycle) to 30 s (120 s cycle). However, the separation by
Cycle B was clearly better with the 40 s cycles than the 120 s cycles
(Table III).

The function performed by cocurrent blowdown can be clearly seen
by a direct comparison of the results of Cycle A and Cycle B. Such
a comparison was made for the 40 s cycles as shown by the performance
curves in Figure 4. For the same N_2 + Ar product purity and product
rate (product rates are given in Table III), the product recovery
of Cycle B was higher than that of Cycle A by approximately 20%for
all runs. This increase in product recovery was a direct consequence
of the cocurrent blowdown step which allowed time for the feed mixture near the feed end to adsorb without being wasted.

A cocurrent blowdown step is also used in many equilibrium separation cycles, however, for a completely different reason (1).

Improvement by Delay Step. Although N_2 is the slow-diffusing component, the isotherms (Figure 1) show a selectivity favoring N_2. The adsorption of N_2, even for small amounts, is detrimental to the kinetic separation. A solution to this problem is to employ a separate delay step (with values at both ends closed), during which the adsorbed N_2 is desorbed but kept within the bed. This amount of N_2 is recovered in the ensuing cycle, thus resulting in a higher product recovery. An experimental demonstration of this important concept is given below.

The experimental data of 4 runs with delay (Cycle C) and 3 runs without delay (Cycle D) are compared in Table IV. At the same N_2 + Ar product rate, both product purity and recovery were higher for Cycle C under a wide range of cycle conditions. Moreover, an interesting pressure rise was observed during the delay step. From the comparison of data in Table IV, the gas desorbed during delay was mostly N_2 + Ar, which increased the product concentration in the voids and consequently the product purity.

In order to better understand the delay step, the N_2 + Ar concentration history of the effluent from cocurrent blowdown (producing step) was measured. Effluent gas samples were taken from a sampling port adjacent to the bed discharge end in different cycles. The data for 4 runs are shown in Table V and Figure 5.

Run 41, using (Cycle D) and Run 44 (Cycle C) cannot be directly compared because conditions were different. However, the superiority of Cycle C over Cycle D is clearly shown by results in Table IV. Run 42 was designed to improve the product purity in Run 41 by collecting less product in cocurrent blowdown. Runs 42 and 44 can be directly compared because they gave nearly the same product throughput rate. Run 44 was clearly better because both product purity and recovery were higher. Run 43 (also using Cycle B which is similar to Cycle D) was a further improvement for Runs 41 and 42, by allowing a longer time for N_2 + Ar desorption to enhance the product purity. Thus Run 43 yielded both high product purity and recovery at nearly the same product rate.

The function performed by delay can be seen in a direct comparison of Runs 43 and 44, both giving the same product throughput rate. The product purity of Run 43 was slightly higher than that in Run 44 (95.5% vs. 95.0%). The major difference was in the product recovery, which was significantly higher in Run 44. This comparison reveals that the column in Run 43 was more thoroughly desorbed than Run 44 at the end of the cycle, but the desorbed N_2 + Ar was partially eluted and discharged, resulting in a lower product recovery. The preceding conclusions are further manifested in the concentration histories shown in Figure 5. Figure 5 shows the concentration of the gas which was essentially the residual gas phase that remained at the end of the preceding cycle. The data indicate that during the last 1/4 cycle, either countercurrent blowdown (Run 43) or delay (Run 44), the desorbed gas was primarily N_2 + Ar. The performance curves of these runs are also compared in Figure 6, showing the same conclusions.

In conclusion, Cycle C is the best cycle for kinetic separation of the air/4A zeolite system. In Cycle C, Steps I and II allow the desorbed in Step III and N_2 + Ar is desorbed in Step IV. Step IV further kept the desorbed N_2 + Ar within the bed as product in the ensuing cycle.

Literature Cited

1. Yang, R.T., Gas Separation by Adsorption Processes, Butterworth: Boston, 1987; Chapters 7 and 8.
2. Keller, II, G.E., Union Carbide Corporation, private communication (1987).
3. Keller, II, G.E., Kuo, C.A., U.S. Patent 4,354,859 (1982).
4. Cramer, R.L., "The AH-64 Nitrogen Inerting Unit," Proceedings Annual SAFE Symp., Asso. Survival and Flight Equipment, p. 198 (1982)
5. Ikels, K.G., Miller, C.W., Thies, C.F., research performed at USAF School of Aerospace Medicine, Brooks AFB, TX (1987).
6. Shin, H.S., Knaebel, K.S. AIChE J., 1987, 33, 654.
7. Ruthven, D.M., Derrah, R.I., J. Chem. Soc. Faraday Trans. I., 1974, 71, 2031.
8. Boniface, H.A., Ruthven, D.M., Chem. Eng. Sci., 1985, 40, 2053.
9. Yeh, Y.T., Dept. of Chem. Eng., SUNY at Buffalo, unpublished results (1987).
10. Yucel, H., Ruthven, D.M., J. Chem. Soc. Faraday Trans. I, 1980, 76, 60.
11. Habgood, H.W., Canad. J. Chem., 1958, 36, 1384.
12. Sarma, P.N., Haynes, H.W., Adv. Chem. Ser. 133, 1974, 205.

Table I. Diffusivities (D) of O_2, N_2 and Ar in 4A Zeolite Crystals

	$D \ (cm^2 s^{-1})$	Sample Used	T,K	Method	Ref.
O_2	3.6×10^{-9}	crystal	303	grav.	7
	3.5×10^{-10}	pellet	303	chrom.	8
	1.18×10^{-9}	crystal	298	grav.	9
N_2	3.8×10^{-11}	crystal	303	grav.	7
	1.2×10^{-9}	crystal	303	grav.	10
	3.8×10^{-10}	crystal	303	grav.	10
	1.3×10^{-10}	pellet	303	grav.	10
	2.8×10^{-12}	pellet	303	vol.	11
	3.4×10^{-13}	pellet	303	chrom.	8
	1.52×10^{-10}	crystal	298	grav.	9
Ar	8×10^{-11}	crystal	303	grav.	7
	10^{-11}	pellet	303	chrom.	12
	9×10^{-13}	pellet	303	chrom.	8

Table II. Cyclic Steady-State Results for Air Separation by Cycle A (Fig. 2) with
Different Cycle Time/Product Rates.

Total Cycle Time, s	20				40				120				360	
Run Number	1	2	3	4	5	6	7	8	9	10	11	12	13	14
N_2 + Ar Product Amount, ℓ STP/cycle	0.45	0.26	0.13	0.05	0.59	0.32	0.19	0.08	0.55	0.35	0.26	0.17	0.50	0.28
Total Feed Amount, ℓ STP/cycle	2.54	2.27	2.13	2.05	3.19	2.94	2.81	2.71	4.05	3.85	3.76	3.67	4.46	4.24
N_2 + Ar Product Rate, ℓ STP/hr/kg 4A	156	90.0	45.0	17.3	102	55.4	32.9	13.8	31.7	20.2	15.0	9.81	9.62	5.38
Total Feed Rate, ℓ STP/hr/kg 4A	848	786	737	710	552	509	486	469	234	222	217	212	85.8	81.5
N_2 + Ar Product Purity, mol %	91.6	94.1	96.2	97.0	92.5	96.3	98.1	99.2	91.6	96.3	97.9	99.4	88.4	93.2
N_2 + Ar Product Recovery, %	21.3	13.6	7.16	3.00	21.6	13.3	8.40	3.70	15.7	11.1	8.57	5.80	12.5	7.80
O_2 Product Purity, mol %	23.8	23.0	22.1	21.5	24.1	23.1	22.4	21.6	23.0	22.7	22.4	22.0	22.2	22.0
O_2 Product Recovery, %	92.7	96.8	98.9	99.7	93.4	98.1	99.4	99.9	94.6	98.4	99.3	99.9	93.6	97.9

P_L = 0 Psig (Low Pressure)

P_H = 25 Psig (High Pressure)

Table III. Cyclic Steady-State Results for Air Separation by Cycle B (Fig. 2) With Different Cycle Time/Product Rates.

Total Cycle Time, s	40				120		
Run Number	15	16	17	18	19	20	21
Cocurrent Blowdown End Pressure, Psig	14.0	16.0	17.5	19.0	15.0	19.0	20.5
N_2 + Ar Product Amount, ℓ STP/cycle	0.50	0.34	0.25	0.13	0.77	0.35	0.18
Feed Amount, ℓ STP/cycle	2.38	2.26	2.32	2.30	3.47	3.43	3.38
N_2 + Ar Product Rate, ℓ STP/hr/kg 4A	86.5	58.9	43.3	22.5	44.4	20.2	10.4
Feed Rate, ℓ STP/hr/kg 4A	412	391	402	398	200	198	195
N_2 + Ar Product Purity, mol %	92.7	95.1	97.0	98.8	89.4	97.0	99.5
N_2 + Ar Product Recovery, %	24.7	18.1	13.2	6.80	25.1	12.5	6.52
O_2 Product Purity, mol %	24.7	23.8	23.2	22.1	24.0	23.0	22.1
O_2 Product Recovery, %	92.7	96.5	98.8	99.5	88.9	98.5	99.9

P_L = 0 Psig (Low Pressure)

P_H = 25 Psig (High Pressure)

Table IV. Cyclic Steady-State Results for Air Separation by Cycle C and Cycle D (Fig. 2) With Different Product Rates.

Total Cycle Time, s	40 (Cycle C)				30 (Cycle C)		
Run Number	22	23	24	25	26	27	28
Cocurrent Blowdown End Pressure, Psig	15.0	16.5	17.8	19.5	16.0	18.0	21.3
N_2 + Ar Product Amount, ℓ STP/cycle	0.51	0.34	0.25	0.14	0.44	0.26	0.11
Feed Amount, ℓ STP/cycle	2.05	1.95	1.96	1.93	2.05	1.99	1.96
N_2 + Ar Product Rate ℓ STP/hr/kg 4A	88.3	58.9	43.3	24.2	102	60.0	25.4
Feed Rate ℓ STP/kg 4A	355	338	339	334	473	459	452
N_2 + Ar Product Purity mol %	91.8	95.0	96.9	98.9	91.4	94.5	97.2
N_2 + Ar Product Recovery %	28.8	21.0	15.6	9.06	24.9	15.8	6.90
O_2 Product Purity mol %	25.2	24.4	23.6	22.6	24.4	23.4	22.1
O_2 Product Recovery %	90.4	96.0	98.1	99.6	91.2	96.5	99.3

P_L = 0 Psig. (Low Pressure)

P_H = 25 Psig. (High Pressure)

Table V. Cyclic Steady-State Air Separation Results for Different Cycles.

Run No.	41	42	43	44
Step I. Feed Pressurization Time, s.	10	10	10	10
Step II. Cocurrent Blowdown Time, s.	10	10	10	10
Step III. Countercurrent Blowdown Time, s.	10	10	20	10
Step IV. Delay Time, s.	No	No	No	10
Total Cycle Time, s.	30	30	40	40
N_2 + Ar Product Amount, ℓ STP/cycle	0.34	0.26	0.34	0.34
Feed Amount, ℓ STP/cycle	2.05	2.01	2.26	1.95
N_2 + Ar Product Throughput, ℓ STP/hr/kg adsorbent	78.5	60.0	58.9	58.9
Feed Throughput, ℓ STP/hr/kg adsorbent	473	464	391	338
N_2 + Ar Product Purity, mol %	93.2	94.8	95.1	95.0
N_2 + Ar Product Recovery, %	19.6	15.5	18.1	21.0
O_2 Product Purity, mol %	23.9	23.3	23.8	24.4
O_2 Product Recovery, %	94.8	96.7	96.5	96.0

P_L = 0 Psig.   (Low Pressure)

P_H = 25 Psig.   (High Pressure)

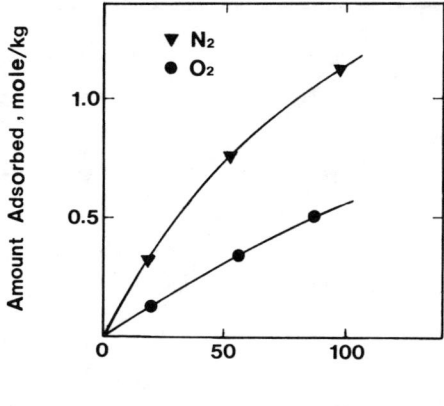

Figure 1.   Adsorption isotherms of O_2 and N_2 in 4A zeolite at 20°C.

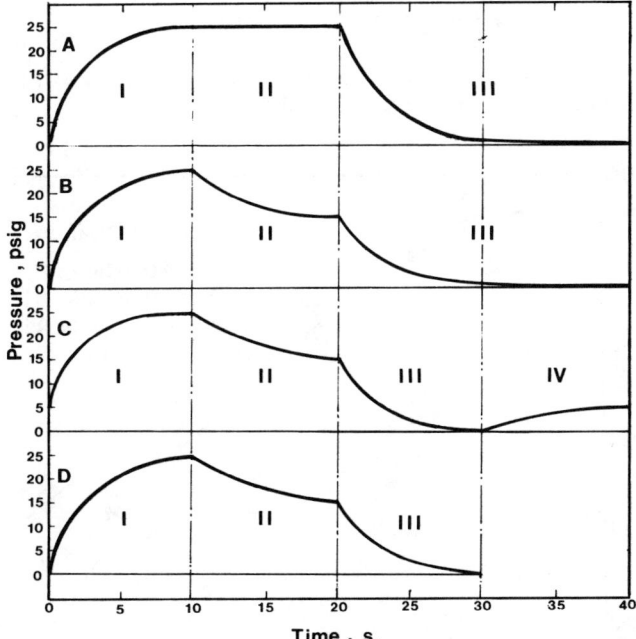

Figure 2.   Pressure histories of the PSA cycles studied for kinetic separation. The slow diffusing component (N_2 + Ar) is the desired product. The cyclic steps are: feed pressurization (Step I); feed or cocurrent blowdown (Step II) (to product N_2 + Ar); countercurrent blowdown (Step III) (to elute O_2); delay (Step IV). For different total cycle times, the pressure histories are changed in proportion with the total time.

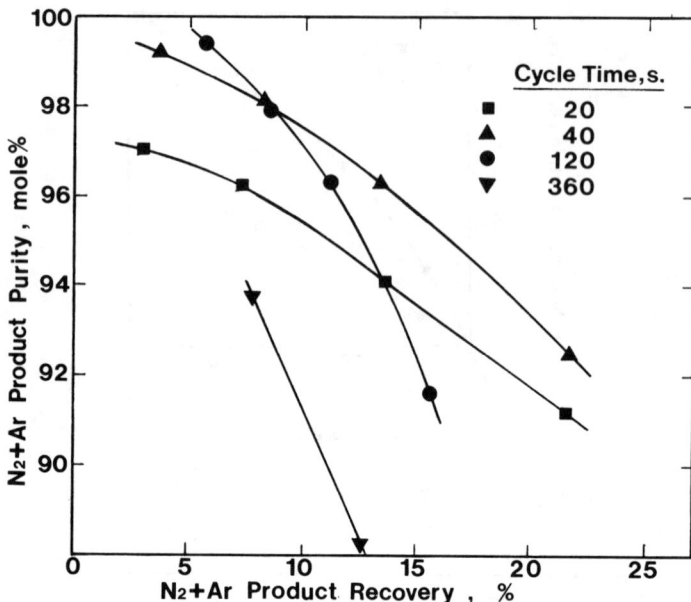

Figure 3. Effect of total cycle time on separation by Cycle A. Detail cycle condition are given in Table II.

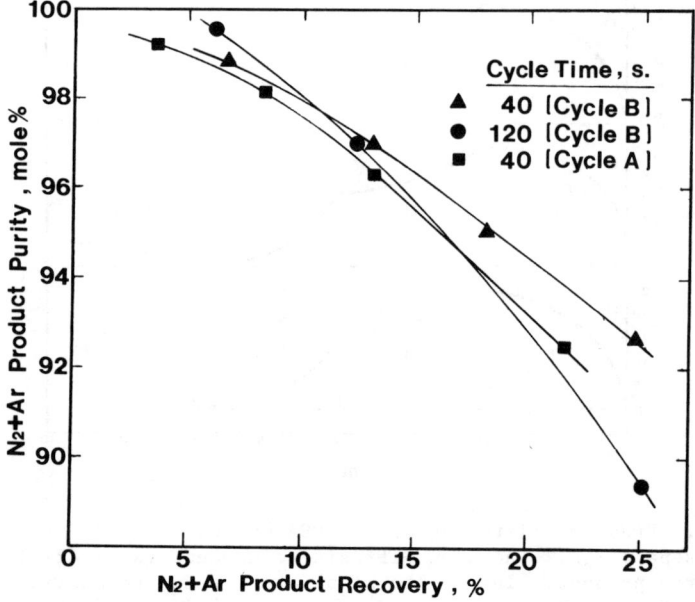

Figure 4. Effect of cocurrent blowdown (used in Cycle B) on separation. Conditions are given in Table III.

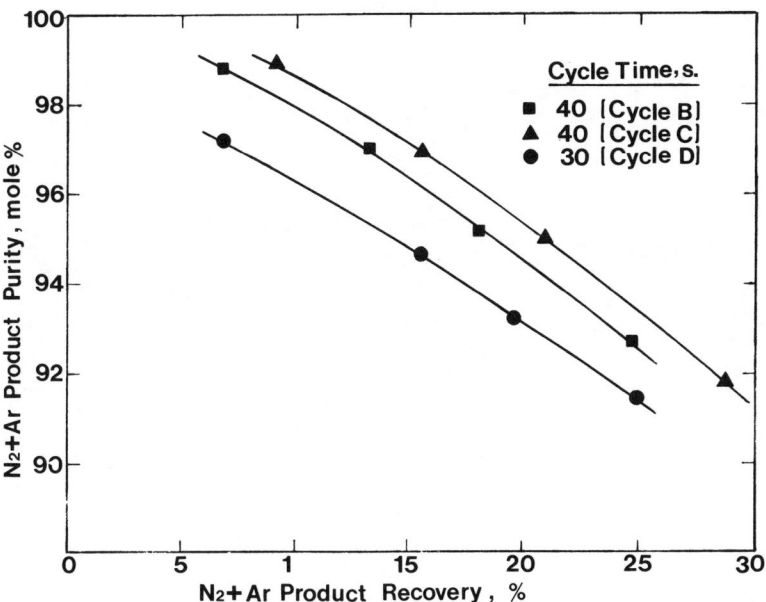

Figure 5. Effluent conceptration histories during cocurrent blowdown. Conditions are given in Table V.

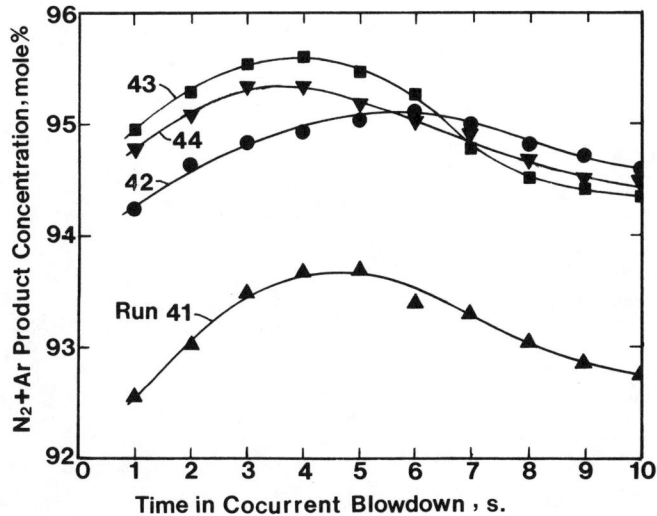

Figure 6. Effect on delay (used in Cycle C) on separation. Conditions given in Table IV.

Chapter 5

Conception of a New Adsorption Process for Purifying Landfill Gas for the Kapiteltal Landfill Site, West Germany

H. Schilling
W. Hinz

- Removal of hydrogen sulphide
- Removal of organic compounds
  (halogenated hydrocarbons)
- Landfill gas analyzing system

On behalf of the landfill-gas utilization facility at the Kapiteltal landfill site, LEYBOLD AG Hanau, West Germany, was commissioned to design and construct the gas purifying plant. Essentially this consists of the stages for the removal of hydrogen sulphide and halogenated hydrocarbons along with the necessary auxiliary systems as well as the gas analyzing system used during the process.

There follows a project definition and a description of the techniques used in the process and the analyzing system.

Introduction

The central Kapiteltal landfill site of the Kaiserslautern Landfill Association, West-Germany, has been in operation since 1976. Its main task is to store, deposit and to treat the rubbish produced by the rural district of Kaiserslautern and the outskirts of the city.

Increased environmental consciousness, as well as higher requirements in the field of waste economy has led to the development of a future-orientated concept of rubbish disposal which incorporates a rubbish-mud-cascade as a main part (see Figure 1).

The job of this plant is, among other things,

the separation of re-usable and reclaimable substances
from household-rubbish as well as the economic utilization of the landfill gas. Due to the enormous power
requirements of the composting and separation plants
of about 800 kW with their very high operating costs,
it became necessary to plan a gas wells and gas ab-
straction system together with combustion engines with
generators for the production of electricity at the
same time.

Due to a high content of halocarbons and sulphur
components found in the landfill gas, which would cor-
rode the engines after a short time, as has been proved
at two landfill gas utilization plants in the Federal
Republic of Germany, where considerable damage occured
in 1982, it was also necessary to install a gas purification plant to protect the engines which will produce
about 1.200 kW. The power produced will be sufficient
to supply the whole installation. Excess electricity
will be fed into the city's electricity grid. (network)

The LEYBOLD AG company, Hanau, Germany was asked
to design and construct the gas purification plant for
the removal of hydrogen sulphide and halogenated hydrocarbons including the necessary auxiliary systems, as
well as the analyzing system to guarantee the safe and
trouble-free working of the downstream engines for the
generation of electricity.

Definition of task

The project LEYBOLD AG is entrusted with consists of
the task of freeing the landfill gas from corrosive
constituents, mainly halogenated hydrocarbons, in order
to ensure the safe and trouble-free operation of the
downstream motors for the generation of electricity.

Furthermore hazardous combustion products, such
as hydrochlorine, hydrofluorine and dioxine, must be
prevented from being emitted to the environment.

This problem was a relatively new one, because
landfill gas, which generally consists of :

> 40 - 60 % methane,
> 30 - 50 % carbon dioxide,
> 0 - 5 % oxygen and
> 0 - 15 % nitrogen,

contains a large number of organic compounds like aromatic and aliphalic hydrocarbons, halogenated hydrocarbons and sulphur compounds (mostly hydrogen sulphide).
Up to 400 traces in landfill gas have been identified.
Some are shown in Figure 2.

Higher contents of chloro- and fluoro-hydrocarbons
and sulphur compounds may cause considerable corrosion
damage as already mentioned.

Therefore the plant had to be specially designed

with respect to the adsorption process used and the selection of materials to ensure
- the safe working of the plant,
- lower maintenance costs and
- low energy consumption.

Further requirements are
- high operational reliability (8000 working hours/
  year minimum)
- high resistance to wear
- long service life of the whole installation
- low risk of explosion
- fully automated operation, i.e. low use of manpower
- optimum utilization of the biogas etc.

The trace constituents must be reduced to the following
values :
Hydrogen sulphide	5 ppmv
Chlorine (total proportion)	20 - 30 mg/m³ (STP)
Fluorine (total proportion)	20 - 30 mg/m³ (STP)

Crude gas specifications

The analysis of the landfill gas before treatment in the
gas purifying plant shows the following composition and
process parameters :

Process parameters

crude gas volume	1000 m³/h (STP)
crude gas pressure	0,92 - 0,96 bar
crude gas temperature	25 - 30°C
relative humidity	100 %

Composition/quality of the crude gas (mean values)

Main components

Methane	CH_4	45 %	by volume
Carbon dioxide	CO_2	38 %	by volume
Nitrogen	N_2	15 %	by volume
Oxygen	O_2	2 %	by volume

Trace components

(Reference "Orientating measurements with the
 Drägerrohr")

Non-halogenated hydrocarbons	Standard mean value:
(Measurements at five well holes)	appr.400 mg/m³ (STP)
Mercaptan (ethyl hydrosulphide)	up to 120 ppm (vol.)
Aldehydes, ketones	up to 150 ppm (vol.)
Hydrogen sulphide	120 - 250 (max) ppm
Ethane and homologues	appr. 50 ppm (vol.)

Analysed trace components
(chlorinated and fluorinated hydrocarbons)

Vinyl chloride	C_2H_3Cl	0 - 16	ppm (vol.)
Vinylidene chloride	$C_2H_2Cl_2$	0 - 1	ppm (vol.)
1.2 trans ethylene dichloride	$C_2 H_2Cl_2$	0 - 1	ppm (vol.)
1.2 cis ethylene dichloride	$C_2H_2Cl_2$	10 - 50	ppm (vol.)
Dichloromethane	CH_2Cl_2	10 -190	ppm (vol.)
Trichloroethylene	C_2HCl_3	5 - 60	ppm (vol.)
Tetrachloroethylene	C_2Cl_4	1 - 15	ppm (vol.)
1.1.1 Trichlorethane	$C_2H_3Cl_3$	0 - 1	ppm (vol.)
F 12	CF_2Cl_2	20 - 30	ppm (vol.)
F 11	CCl_3F	2 - 10	ppm (vol.)

With reference to the allowable proportions of
chlorine and fluorine the above values have
the following meaning:

1000 - 1300 mg Chlorine/m³ (STP) crude gas
  70 - 100 mg Fluorine /m³ (STP) crude gas

Purification process

The plant, which is able to creat 1000 m³/h (STP) of
landfill gas, consists of the following process stages:
- Condensation of untreated landfill gas (Dehydration)
- Removal of hydrogen sulphide
- Removal of organic compounds
- Environmentally safe disposal of the trace components
  produced during regeneration of the plant
- Analyzing system for the measurement of oxygen, me-
  thane, hydrogen sulphide, humidity, chlorine and
  fluorine.

Different types of activated charcoal have been used for
the desulphurization as well as the purification of
landfill gas.
    The process technique chosen by LEYBOLD AG is shown
in Figure 3.

Removal of H_2S

The removal of hydrogen sulphide and organic sulphur
compounds has been achieved for decades by means of
activated charcoal many synthesis processes in industrial chemistry would not have been possible without
it (1).
    The desulphurization process of landfill gas chosen
by LEYBOLD-AG is known as the fixed bed process using
wide-pore iodized activated charcoal both as adsorber
and catalyst. This process utilizes the catalytic properties of iodine-impregnated activated charcoal. In

the so-called adsorption catalysis the catalytic effects
of activated charcoal are used to reduce the H_2S to
sulphur in the presence of oxygen.

Catalytic oxidation of H_2S

$$2\ H_2S + O_2 \underline{\text{ activated charcoal }} \quad 1/4\ S_8 + 2\ H_2O$$

The elemental sulphur generated in this process is adsorbed while the reaction product, i.e. water, is desorbed by the surface of the catalyst (2).

After this stage the residual contents of hydrogen
sulphide lie below 1 mg/m³.

The loads of sulphur may reach 100 % by weight.

In most cases the oxygen required for the oxida-
tion will be present in the landfill gas; if the content should prove to be insufficient, a small and
controlled amount of air is fed in automatically.

Two fixed bed adsorbers designed for a service
life of up to six months are fitted. Once the hydrogen
sulphide contents at the outlet following the H_2S-re-
moval reach a value of 5 ppm, the second adsorber is
activated. Then the first adsorber may first be flushed
with nitrogen and subsequently emptied to be refilled
with fresh activated charcoal.

After being inertized and methanized once again,
the adsorber is ready for another process cycle.
The loaded charcoal will be discarded.

Removal of organic contamination

With reference to the concepts for the recovery of solvents, a new process for the removal of organic conta-
minations has been developed. In solvent-recovery
plants unary or binary mixtures in a concentration
range between 1.000 - 10.000 mg/m³ have been treated.

In contrast to solvent polluted air, landfill gas
is a polynary mixture with contaminant-concentrations
in a range between 1 - 500 mg/m³.

The quantity of contaminants is not limited to halogenated hydrocarbons. Landfill gas contains aromates,
alkanes and a great many of other contaminants as well.

Because of their different boiling temperatures,
the contaminants are divided in to high-, medium- and
low boiling traceconstituents. Figure 4 shows three
typical examples of components found in landfill gas
and their adsorption balances on activated charcoal.

Volatile contaminants could be displaced by high
boiling constituents.

Adsorption displacement has to be considered
during the design of the process.

Therefore a variety of activated charcoal types
has been used in order to remove adsorptively and selectively the high-, medium- and low boiling consti-

tuents.

Design and dimensioning of this stage are based on the minimum/maximum analysis results of the landfill gas to be processed.

In addition to the above, the adsorption balances, the desorption energies and the co-adsorption of the different trace components must be taken into consideration.

High gas humidity in combination with a low concentration of halocarbons will influence the adsorption capacity of the activated charcoal. Optimized desorption by increased steaming and a sufficient drying of the charcoal must be planned for.

It has been found out, that three times the quantity of steam with a double steaming time brought a three fold adsorption time.

In addition to that, an optimized drying cycle has brought a five fold improvement in final purity.

The charging pressures used at the Kapiteltal gas purification plant are about p = 1,3 bar; at plants with downstream pressure swing adsorption units for the enrichment of methane such pressures may reach up to p = 9 bar.

Contrary to the conventional equipment for the recovery of solvents, the regeneration is not conducted in one stage but in two stages using steam and hot gas, the latter being purified landfill gas. Following the hot gas desorption cycle and subsequent drying, the adsorber is cooled and flushed with the same gas.

The specific techniques of this process will ensure that the organic constituents contained in the gas may be extracted in a concentrated form and separated with a high degree of accuracy. Furthermore, an optimum desorption of higher-boiling constituents is also ensured (3).

Any non-condensable trace components are channelled, together with the flush gas, into the waste gas chimney, a procedure in line with the latest air pollution regulations (TA-Luft).

The condensable constituents are, together with the overall condensate, collected in a separating recipient, separated there and routed into two additional containers.

In this way a solvent-enriched phase and a water-enriched phase is obtained. The solvent-enriched phase is subjected to specific treatment while the water-enriched phase, together with other condensates collected at the plant, is sent to a condensate cleaning unit.

Overall plant layout

The following schematic flow diagram of the gas puri-

fication plant (Figure 5) shows, that the plant is
equipped with a number of auxiliary installations
and subsystems. The sections mentioned below represent the most important parts of the whole plant
which was designed for fully automated operation:

- crude gas filter
- crude gas water condenser with demister
- crude gas heater
- crude gas blower
- hydrogensulphide separator with aftercooler
- HML-system
- steam generator with automatic sludge drain
  and feedwater conditioning
- refrigerating system (split system)
- analyzing system.

Program sequence (Figure 6)

In its schematic version the program sequence diagram
shows the steps required for putting the plant into
operation. After the subsystems have been activated
and the respective rated values have been reached,
the entire plant is flushed with nitrogen and methanized. During this phase the analyzing system
works as the process control system.

When in the methanizing mode the plant is run
in a cycle with the flush gas output via a flare
until the gas in the system has reached the quality
required for powering the engines. At this point
the engines may be started.

All subsystems as well as the gas quality are
monitored automatically. In the event of a fault
the plant will assume the correct stand-by mode.

Explosion and safety precautions

The plant has been built into three separate rooms.
The switch panel and the analysis equipment as well
as the steam generator and the refrigeration unit
has been built into two rooms which are declared
as non hazardous areas. The gas purification plant,
has been built into a room declared as Zone 1 which
means that occasionally a dangerous, potentially explosive atmosphere can be expected. The safety regulations according to Ex-RL have to be fulfilled.

For this reason the formation of an atmosphere
with high quantities of explosive mixtures must be
avoided. Primary precautions are among others :

- Avoidance of inflammable material which
  could create an explosive atmosphere
- Limitation of the quantity or concentration

of inflammable materials
- Powered ventilation
- Inertisation
- Checking of concentration with automatic
  release of protective safety precautions,
  i.e. use of gas warning equipment.

These requirements have led to the installation of

- a gas analysing system
- gas warning equipment in eight different places
- a powered room ventilation system as well as the
  usage of approved parts according to the Ex-RL-
  regulations.

Due to these precautionary features the entire plant
possesses a high degree of operational safety.
Once, for example, the safety limit values (e.g. 8% O_2
or 20% methane) are exceeded, the safety precaution
functions operate, which will keep the plant in a state
of safety (see literature reference 4), or return it to
such.
Detailed information may be obtained in section 7
"Analyzing System".

Corrosion protection

All parts of the plant except of the HML-system have
been manufactured in Stainless Steel.
For the HML-system different materials have been
used. Because of the formation of HCL and HF (hydro-
chloric and hydrofluoric acid) during regeneration of
the adsorbers, it has been especially protected. The
adsorbers are enamel coated. The piping of the regeneration cycle has Teflonlining. The two heat exchangers are made of a special type of Stainless Steel
and of graphite respectively.
Furthermore special gaskets made of Viton or
PTFE has been used.

Specifications

Wastewater
Condensate yield (appr.)	1,12 m³/day
High/medium/low-boiling yield	30 - 58 kg/day

Operating materials/utilities
Electric power supply (installed)	360 kW
" " consumption	90 kW
Specific " "	0,09 kWh/m³ (STP)
Water consumption	0,5 m³/day
Compressed air cons. (5-7 bar)	2,4 m³/day

Process gas (purified) cons.   60 m³/day
                                           $\triangleq$ 0,25 %

Activated charcoal consumption   5 m³/6 months
                                           at 250 ppm H_2S

Nitrogen consumption   1 m³/day

Space requirement
    Total space requirement   600 m³
    Control and analyzing room
    system   24 m³
    Auxiliary equipment   95 m³
    Total floor space required   125 m³

The analyzing system

The entire gas purification plant is monitored and controlled by means of measurements taken of selected components at various points of the process flow. In the present plant the following gas components are measured:

- Oxygen
- Methane
- Water vapour
- Hydrogen sulphide

Three different physical principles are employed in order to determine analytically the presence and volume of these four gas components.

The measurement of oxygen utilizes the paramagnetic properties of the oxygen molecule. Due to this paramagnetism, oxygen will, when subjected to an inhomogeneous magnetic field, be drawn into the area of higher field intensities. Figure 7 shows the schematic representation of such analyzer.

The measuring cell of Oxynos 1 contains a small dumbbel-shaped object consisting of nitrogen-filled quartz balls, one at either end of it. This object is suspended horizontally by means of a thin tight platinum band. A small mirror has been attached to the center of said dumbbellshaped object, the latter being surrounded by a coil.

The measuring cell itself is surrounded by permanent magnets. These magnets generate a strongly inhomogeneous magnetic field. Once oxygen molecules are carried into the measuring cell as part of the gas to be measured, these oxygen molecules will be deflected into the area of the largest field strength. This deflection exercises different forces on either one of the quartz balls, thereby turning the dumbbell-shaped object out of its zero position. As soon as the mirror attached to the object is displaced out of its zero position, the signal of a detector will generate a current subsequently to be amplified.

This current is sent through the surrounding coil creating a homogeneous magnetic field which forces the dumbbell-shaped object back into its zero position. Thus the strength of such current is directly porportionate to the magnitude of the oxygen concentration present in the cell; the current is measured with the help of a galvanometer and displayed as desired.

The measurement of the components methane and water vapour is achieved by means of a two-channel non-dispersive infrared gas analyzer of the type "Binos". Figure 8 shows a schematic rendition of the functional principle of the Binos gas analyzer:

Infrare-active gases absorb heat radiation of wave-lengths being specific for the type of gas in question.

A heating spiral generates the necessary IR-radiation. This IR-radiation is split into two beams of identical intensity and sent through the analyzing cell, the latter being respectively divided into a sampling or measuring side and a reference or comparison side. A subsequent filtering section of the cell eliminates the unwanted portions of the radiation spectrum and provides adaptation for the clear aperature of the detector. A light chopper wheel in a gastight housing is running horizontally below the filtering section of the cell. The shape of the chopper has been carefully designed to ensure a high degree of stability in sensitivity. The chopper wheel is driven by means of two eddy-current magnets.

The pneumatic radiation detector selectivity captures the radiation from the sampling and the reference side as released alternatingly by the chopper wheel and converts it into a voltage being proportionate to the radiation intensity.

In this case the infrared measuring technique was chosen for the water vapour, too, because this method offered hardly any problems regarding corrosion and contamination. Furthermore, this measuring technique is very accurate and reproducible.

A semiconductor-type sensor is used to measure the hydrogen sulphide contents. Sensors of the semiconductor-type change their conductivity caused by a charge exchange at their surface: this change of conductivity is evaluated electronically. Some precautionary measures are indicated for the use of such sensors in landfill gas technology to have work with sufficient accuracy. As for the rest, measuring with semiconductor-type sensors is the most economic method available on the market today.

Now the reasons for installing the analyzers at their specific locations. Reference is made to Figure 8 (Process flow diagram of the gas purification plant). Before entering the plant, the landfill gas is measured by a continuously operating analyzer for its contents of oxygen. An excessive content of oxygen involves the risk of reaching the threshold of an explosion.

Within the meaning of the provisions for explosion protection the existing roots blower is considered to be the external, source of ignition. Furthermore, the O_2-content may be used to determine the rate of density. Once the O_2-content exceeds a preset upper limit, the system is shut down automatically.

The next measurement of oxygen serves the purpose of checking upper and lower limit values. If not enough oxygen is present, the H_2S-conversion is in jeopardy. Hence, when an oxygen deficiency occurs, the process control will automatically activate a blower to pump air into the system. This additive process may be controlled by a second measuring point, if required.

Another measuring point is activated at the inertizing phase of the process where it controlls the necessary O_2-content. Again, this is a preventive measure for explosion protection.

Methane is measured at three points throughout the plant. One measurement at the gas input controls the CH_4-content and may trigger a shut-down of the plant once the content falls below a lower limit value, as the operation of the gas motors requires a certain minimum content of CH_4.

The measurement at the gas output helps monitoring and supervising the entire system and may also be used to determine the energy value of the landfill gas.

The third measuring point has already been mentioned before; it controls the inertizing process.

The measuring point "water vapour" registers the state of the gas after leaving the heat exchanger. Due to reasons based on the process technology, the humidity of the landfill gas must not fall below the value measured at this point. The measurement of H_2S is provided downstream the adsorber; it controls the efficiency of the latter and informs the operating staff when it is time to switch over from one adsorber to the other.

The environment of the entire plant is controlled continuously by an independent explosion protection system which will detect any escaping landfill gas.

The entire analyzing system is monitored and controlled by means of a freely programmable control system. Any fault reports are forwarded automatically to the CPU (Central Processing Unit) and also displayed on site.

Literature Cited

1.   Von Kienle, Dr.H.; Bäder, Dr.E. <u>Aktivkohle und</u>

ihre industrielle Anwendung; Ferdinand Enke Verlag, Stuttgart 1980.

2.   Henning, K. D.; Klein, J.; Knoblauch, K.; Schwefelwasserstoff-Entfernung aus Biogas mit einem Aktivkohleverfahren. Special edition of "GWF-Gas/Erdgas" 126 (1985), issue 1, p. 19 - 24.

3.   Schilling, H., Leybold AG. Henning, K. D., Bergbauforschung mbH. Erdgasgewinnung aus Deponie- und Faulgasen. (Special edition by Leybold AG).

4.   Redeker, T.; Schampel, K.; Explosionsschutz an Anlagen zur Absaugung, Aufbereitung und Nutzung von Biogas aus Mülldeponien. In: Deponietechnik Heute und Morgen, volume 19. Erich Schmidt Verlag, Bielefeld 1985.

RUBBISH MUD CASCADE KAPITELTAL WEST-GERMANY

MÜLL- KLÄRSCHLAMM CASCADE KAPITELTAL

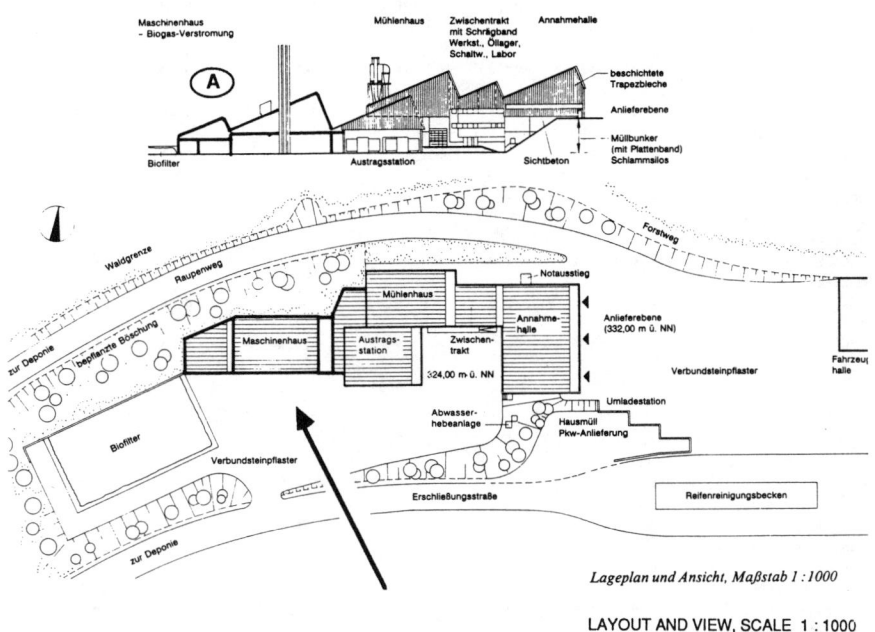

Lageplan und Ansicht, Maßstab 1 : 1000

LAYOUT AND VIEW, SCALE 1 : 1000

(A) POWERHOUSE
LANDFILL GAS PURIFICATION AND
CURRENT GENERATION

Fig. 1

Halogenated Hydrocarbons	Asslar	Ahrenshöft	Braunschweig	Karlsruhe West	Kaiserslautern Kapiteltal	Deponie Wannsee	Fresh Kills USA
				(mg/m³)			
Dichlordifluormethane (Freon 12) CCl_2F_2	131,0	103,1	36,2	25,8	162	0,35	22,1
Trichlorfluormethane (Freon 11) CCl_3F	26,0	33,1	47,8	11,9	61,4	0,03	-
Dichlormethane CH_2Cl_2	730,0	-	10,9	0,012	720,1	0,49	0,34
Trichlormethane $CHCl_3$	0,5	-	10,7	-	-	0,01	
Carbontetrachlorine CCl_4	-	-	-	0,001	-	0,008	-
1.2.2-Trifluortri- chlorethane $C_2Cl_3F_2$	1,4	1,7	1,7	-	-	-	-
Chlorethane C_2H_5Cl	-	-	9,5	-	-	0,26	-
1.1-Dichlorethane $C_2H_4Cl_2$	-	15,0	-	-	220,0	1,34	-
1.1.1-Trichlorethane $C_2H_3Cl_3$	2,9	2,9	0,6	0,017	5,9	-	-
Trichlorethylene C_2HCl_3	47,9	12,3	31,7	0,006	352,3	2,0	1,8
Tetrachlorethylen C_2Cl_4	55,8	14,1	83,6	0,006	65,0	1,3	1,5
Vinylchloride C_2H_3Cl	11,1	-	-	-	44,6	0,27	0,6
Vinylidenchloride $C_2H_2Cl_2$	0,5	-	7,8	-	4,33	-	3,9
cis-Dichlorethylene $C_2H_4Cl_2$	49,5	3,9	294,4	-	-	0,08	-
trans-Dichlorethylene $C_2H_4Cl_2$	0,8	-	-	-	-	-	-
Total-Chlorine	842,0	125,1	401,9	24,5	1265,0	10,6	25,1
Total-Fluorine	45,1	37,5	18,5	9,8	59,3	0,6	11,9

Figure : 2   Results of halogenated contents of different landfill gases (acc. literature)

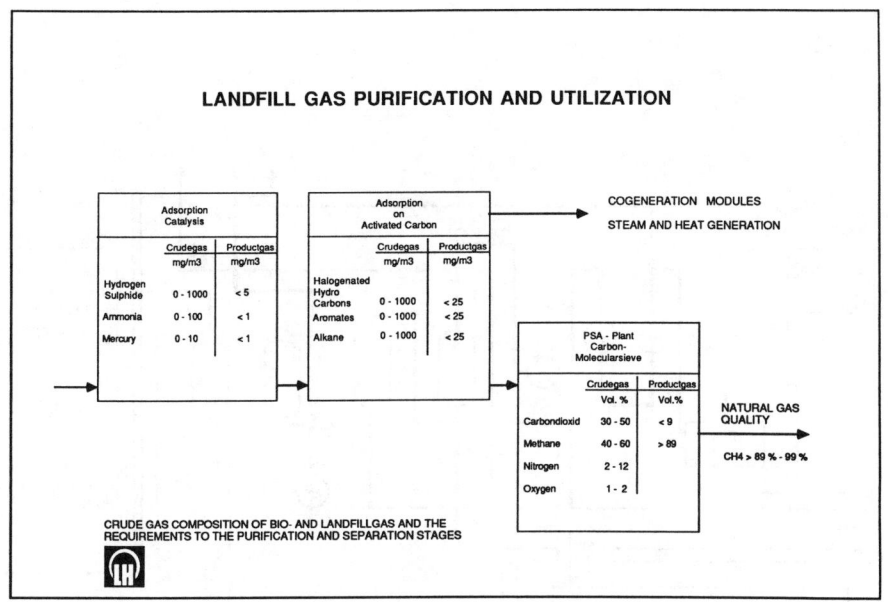

Fig. 3

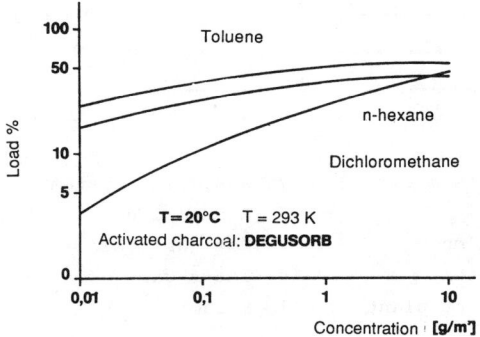

**Adsorption balances of organic vapours
on activated charcoal**

Fig. 4

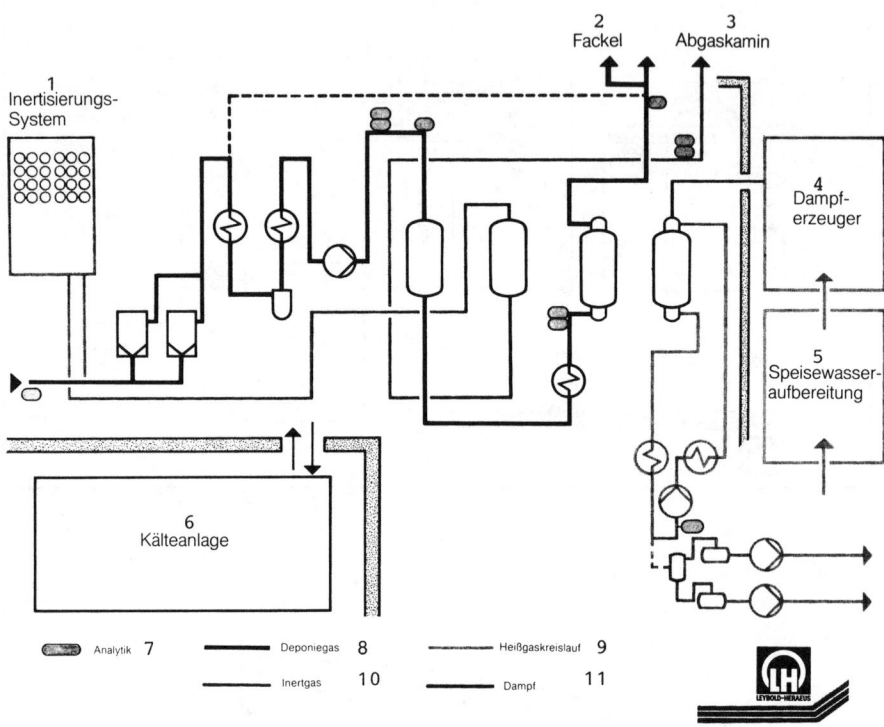

1   Inertisation system   7   Analysing system
2   Flare   8   Landfill gas
3   Waste gas chimney   9   Hot gas cycle
4   Steam generator   10   Inert gas
5   Water treatment plant   11   Steam
6   Chilling unit

Fig. 5

WORKING   PROGRAMME

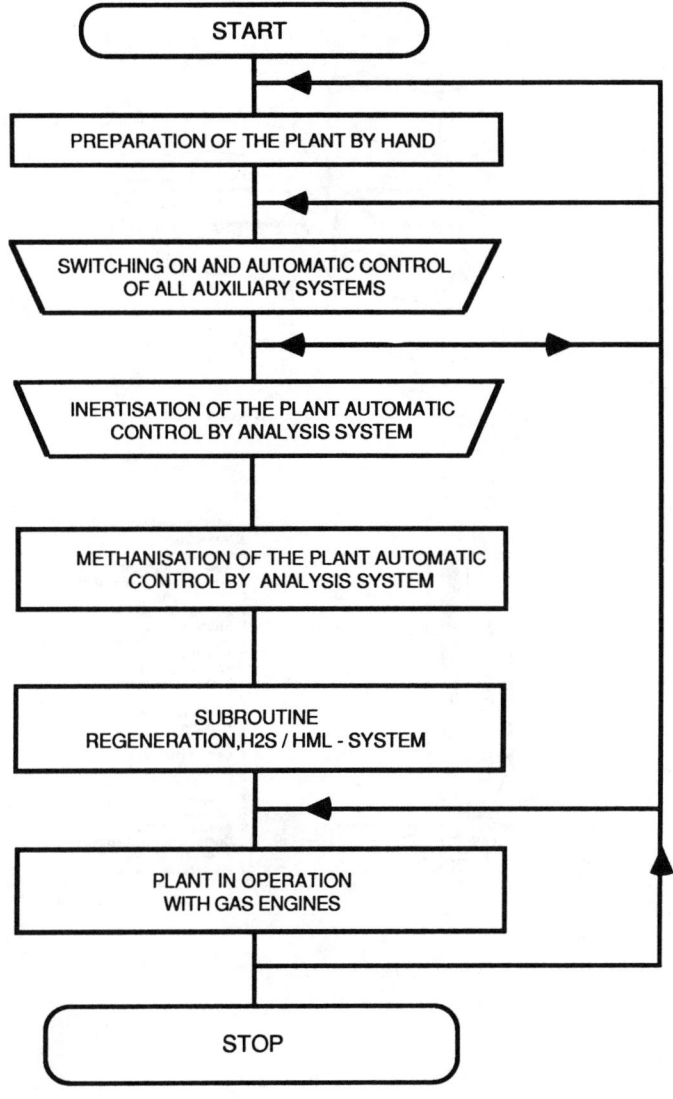

Fig. 6

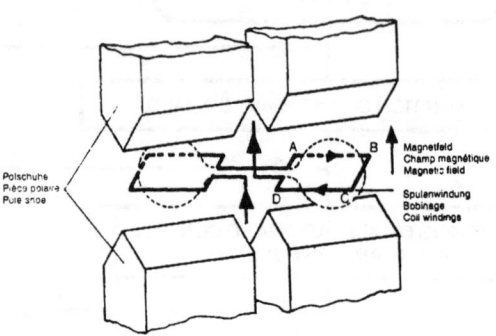

Fig. 7 : Diagrammatic view of Oxynos 1

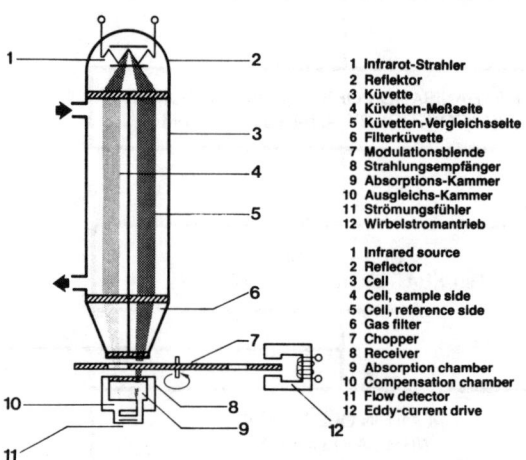

1 Infrarot-Strahler
2 Reflektor
3 Küvette
4 Küvetten-Meßseite
5 Küvetten-Vergleichsseite
6 Filterküvette
7 Modulationsblende
8 Strahlungsempfänger
9 Absorptions-Kammer
10 Ausgleichs-Kammer
11 Strömungsfühler
12 Wirbelstromantrieb

1 Infrared source
2 Reflector
3 Cell
4 Cell, sample side
5 Cell, reference side
6 Gas filter
7 Chopper
8 Receiver
9 Absorption chamber
10 Compensation chamber
11 Flow detector
12 Eddy-current drive

Fig. 8   Vertical section of BINOS IR

Chapter 6

Sizing of Vacuum Pumps for Desorption in PSA Systems

Heinrich Amlinger

The sizing of vacuum pumps for desorption has to be
based on the desorption characteristics of the adsorbent.
This desorption characteristic differs from the type
of application for adsorption systems. Thermal desorption is different from isothermal desorption and
desorption for pressure swing systems. The paper will
show, how the suction capacity of a vacuum pump system
can be calculated for different desorption requirements.
The energy consumption is most important for adsorption
processes and influences the plant economics. It will
be shown, how vacuum pump systems can be optimized
for adsorption processes, in order to reduce energy
consumption. For different vacuum desorption pump
systems, the course of evacuation time and the energy
consumption will be demonstrated. The energy consumption will be minimized, if the vacuum pumps are
optimized by multi-stage pumps.

General Consideration

Sizing of a suitable vacuum pump for desorption is essentially
influenced by the desorption characteristics of the molecular sieve.
    This desorption characteristic is, in turn, greatly influenced
by the adsorption of the adsorption energy, respectively. When
looking at a certain particle of a molecular sieve, it becomes
evident that the so-called physical adsorption takes place in the
potential minimum of the centre of attraction, or force, of the
particle's surface.
    When the adsorbed gas molecule reaches the potential minimum, a
form of energy, i.e. the so-called adsorption energy q_a, is set free.
Contrary hereto only such molecules as feature a certain energy or
velocity will be able to leave the potential minimum again. The
mean residence time $\bar{t}$ in the potential trough is calculated according
to the following equation:

$$\bar{t} = \frac{1}{\nu} e^{q_a/RT}$$ (1)

where $1/\nu$ is the time of one lattice vibration of the solid atoms, $1/\nu = 10^{12}$ to 10^{14} s. The adsorption energy of CO_2 with regard to carbon is approximately 7.73×10^5 J/kg. In this case the mean residence time of a CO_2 molecule will be approximately 7.35×10^{-6} s.

If, in the case of vacuum desorption, one assumes that every gas molecule which is desorbed in the vacuum by the surface of the molecular sieve particle will not return to the particle surface, but will be pumped off, the decrease in the surface concentration of the adsorbed gases within the time dt will be:

$$db = -\frac{b}{t} \cdot dt$$

Assuming as an initial condition, that at the time t = o the surface concentration is $b = b_o$, the integration of the above equation will result in:

$$db = b_o \int_{t}^{t} \frac{dt}{t}$$ (2)

$$b = b_o \ln \frac{t}{t}$$

where b_o = initial concentration and b = residual concentration.

As can be seen from equation (1), the mean residence time $\bar{t}$ will increase with decreasing temperature, and will decrease with increasing temperature: an increase in temperature accelerates desorption, whereas a decrease in temperature slows desorption down; equation (2) taking into account the mean residence time $\bar{t}$.

Equation (2) was derived under the assumption of an overall uniform surface. However, this cannot be assumed for every particle of the molecular sieve; so, in fact, equation (2) should rather be written as follows:

$$b = b_o \ln \frac{t}{t} + b_1 \ln \frac{t}{t_1} + \dots$$ (2a)

Owing to the concurrent action of different exponential functions it is absolutely possible that the course of the desorption curve is different to that of an exponential curve. It is possible that a combination of a multitude of exponential curves leads to an approximately linear course of curve in wide ranges.

As can be seen from equation (2), it is extraordinarily difficult to predetermine desorption courses on a merely calculatory basis, i.e. without carrying out any experiments. It is at best possible to forecast certain trends.

The course of desorption over time is not necessarily identical with the course of evacuation over time in the case of desorption with a vacuum pump (see Fig. 1). A coincidence of these two curves would be pure chance. However, the mean desorption gas quantity per unit of time and the mean gas quantity pumped off by the vacuum pump per unit of time have to be identical, since otherwise the balance of the adsorption gas quantity would not correspond to the balance of the desorption gas quantity.

Balancing of Gas Quantity to be Pumped Off

A vacuum pump does not only have to pump off the desorption gas quantities, but also the adsorber's gap and clearance columns (equation (3)). It is possible to reduce the gas and clearance volumes to be pumped off to a considerable degree by using a corresponding process control.

In the case of adsorption installations in which a pressure compensation preceeds the desorption operation, it is possible to create a slight vacuum in the adsorber to be desorbed by means of an appropriate reduction of the adsorption pressure compensation. Owing to the lower pressure occurring in the adsorber, the gas and clearance gas quantity as well as the charge gas quantity and the energy consumption are correspondingly decreased. In the case of alternating pressure installations with adsorption pressures which leave – after pressure compensation, but prior to evacuation – in the adsorber a pressure greater than atmosphere, the pressure balance above atmosphere is released prior to evacuation.

In the case of installations of this type the charge gas quantity is greater than in the case of other installations in which atmospheric pressure is already achieved after pressure compensation. This has to be taken into account, when balancing the gas quantity to be pumped off by the vacuum pump.

Balancing of the gas quantities of a PSA installation with vacuum regeneration:

Total off-gas quantity:

$$V_1 = V_2 - V_3 = V_4 + V_5 \frac{P_d}{P_o} \quad [\frac{m^3 n}{t} \quad or \quad \frac{m^3 n}{cycle}] \tag{3}$$

To be pumped off by the vacuum pump:

$$V_o = V_4 + V_5 = V_1 - V_5(\frac{P_d}{P_o} - 1) \quad [\frac{m^3 n}{t} \quad or \quad \frac{m^3 n}{cycle}] \tag{4}$$

where V_o = gas quantity to be pumped off by the vacuum pump, V_1 = off gas quantity, V_2 = charge gas quantity, V_3 = product gas quantity, V_4 = adsorbed gas/desorbed gas, V_5 = gas and clearance volumes of the adsorber, P_d = pressure compensation pressure, P_o = atmospheric pressure on site, and $m^3 n$ denotes m^3 NTP.

In addition, the gas quantity V_o to be pumped off by the vacuum pump has to be corrected according to the intake temperature and intake pressure.

$$V = \frac{V_o P_n T_1}{P_o T_n} \quad [\frac{m^3}{t} \quad or \quad \frac{m^3}{cycle}] \tag{5}$$

where *P_n = 1013 (mbar) standard pressure, T_n = 273 K, and T_1 = temperature in intake condition (K). Every pressure above atmosphere is released prior to evacuation.

Calculation Model for Determination of Suction Capacity of the Vacuum Pump

At the intake nozzle of the vacuum pump, the gas volume V changes according to the decreasing pressure P, according to Boyle-Mariotte's law at T_1 = constant.

$$V_v = \frac{V \; P_o}{P} = \frac{V}{t} \cdot \frac{P_o}{P}$$

V_v = S = Gas quantity or suction quantity, resp. for the corresponding vacuum pressure (m^3/t)

P = vacuum pressure (mbar)

$$\Delta S = \frac{V \cdot \Delta P}{P} = \frac{V}{t} \cdot \frac{\Delta P}{P} \quad [\frac{m^3}{t}] \tag{6}$$

$$dS = \frac{V}{t} \cdot \frac{dP}{P}$$

Equation (6) shows that the suction capacity of the vacuum pump has to increase by ΔS within each constant time interval as the pressure P decreases. Fig. 2 shows this reversed proportionality S = f (p).

$$\text{The term } V \cdot P_o = Q[\frac{mbar \; m^3}{t}]$$

is referred to as mass flow which is constant with regard to each evacuation interval. In the double-logarithmic graticule the mass flow $Q = V \cdot P_o$ forms a straight line under an angle of 45°. This form of representation is used for the calculation of multi-stage vacuum pumps in order to determine for the respective operating point from one stage the size of the next one. (Q = constant).

In contrast to the representation in Fig. 2, a vacuum pump cannot permanently increase the suction capacity while the ultimate pressure drops; it can at best keep it constant.

The result of the integration of equation (6) is the mean required constant suction capacity of vacuum pump in the course of the evacuation of a container from pressure P_o to P within the time interval t

$$\int dS = -\frac{V}{t} \int_{P}^{P_o} \frac{dP}{P}$$

$$S = \frac{V}{t} \cdot \ln \frac{P_o}{P} \tag{7}$$

If one uses this equation for the determination of the suction capacity of a vacuum pump for desorption purposes, this set-up will in almost every case be too optimistic.

The model of a simple container evacuation leads to suction capacity values which are too small, since the desorption of the adsorbent requires the conveying of a larger gas quantity at a lower pressure.

However, if the container model is used for the calculation of a partial section, the values obtained for the determination of the requires suction capacity for desorption will be quite accurate.

When evacuating a container, or adsorber, in partial sections from the respective pressure P_1 to P_2, the total of all the gas subquantities $\bar{v}$ will again be the total of the gas quantity V to be pumped off.

Gas subquantity $\quad \bar{V} = V \dfrac{(P_1-P_2)}{P_o} \quad (m^3)$

If pumping off of the gas subquantity is carried out within the time interval (t_1-t_2),

then $\qquad \dfrac{\bar{V}}{(t_1-t_2)} = \dfrac{V}{t} \dfrac{(P_1-P_2)}{P_o} \quad \dfrac{(m^3)}{t}$

or $\qquad d\dfrac{V}{t} = dS = \dfrac{P_o}{(t_1-t_2)(P_1-P_2)} \cdot \dfrac{dP}{P}$

The solution will in principle lead to equation (7) again, however, for the calculation of a container's or adsorber's partial section.

$$\int dS = -\dfrac{\bar{V} \cdot P_o}{(t_1-t_2)(P_2-P_2)} \int_{P_2}^{P_1} \dfrac{dp}{P}$$

$$S = \dfrac{\bar{V} \cdot P_o}{(t_1-t_2)(P_1-P_2)} \ln \dfrac{P_1}{P_2}$$

or for constant S:

$$\bar{V} = \dfrac{S \cdot (t_1-t_2)(P_1-P_2)}{P_o \cdot \ln \dfrac{P_1}{P_2}} \qquad\qquad (8)$$

where P_1 = pressure at the beginning of the interval, P_2 = pressure at the end of the interval, t_1 = time at the beginning of the interval, and t_2 = time at the end of the interval.

$\bar{V}$ is the mean gas subquantity (m^3/cycle), occuring at a constant suction capacity denoted by S m^3/s for a certain pressure and time interval.

Equation (8) can be derived in a different way: An assumed suction capacity (S_o) is used to determine the mass flow Q for each interval.

$$Q_{1,2} = \bar{V} \cdot Po_{1,2} = -S_o \int_1^2 P(t) \cdot dt$$

$$P(t) = P_1 \cdot \exp \dfrac{-t_2-t_1}{\tau}$$

with the fixed points (P_1 and P_2 known):

$$P_2(t) = P_1 \cdot \exp \dfrac{-t_2-t_1}{\tau_{12}}, \quad \tau_{12} \text{ to be determined by:}$$

$$\tau_{12} = \frac{t_2 - t_1}{\ln \dfrac{P_1}{P_2}}$$

$$Q_{1,2} = -S_o P_1 \int_1^2 e^{-t-t_1/\tau_{12}}$$

$$Q_{1,2} = S_o P_1 \tau_{12}(1-e^{-(t_2-t_1)/\tau_{12}}$$

$$Q_{1,2} = S_o \tau_{12}(P_1-P_2)$$

$$Q_{12} = \overline{V}Po_{12} = \frac{S_o(t_2-t_1)(P_1-P_2)}{\ln \dfrac{P_1}{P_2}}$$

$$\overline{V} = \frac{S_o(t_1-t_2)(P_1-P_2)}{P_o \ln \dfrac{P_1}{P_2}}$$

As one can see, the result of the solution is again equation (8).

For the calculation of the suction capacity, the evacuation curve or desorption curve, respectively, is broken down to individual partial intervals. Using equation (8), V_1-n is determined for each partial interval under the assumption of S_o = constant.

With equation (7), (S_o) can at first be determined as 0-approximation. Sum V_1-n is formed. Then the actual suction capacity S_s is calculated.

$$S_s = S_o \frac{V}{\Sigma(\overline{V}_1-n)} \geq S_o \qquad (9)$$

V is determined according to equations (4) and (5). It is a prerequisite for the model described above that the adsorbent capacity is fully used with a given product gas purity, or respectively, that this adsorbent capacity is large.

With O_2 and N_2 on zeolites and carbon molecular sieves, the adsorbent capacity is between 3.5 and 5 m^3 gas/m^3 sorbent, and with CH_4 adsorption on activated charcoal, up to 12 m^3 gas/m^3 sorbent. These values refer to a relatively high product gas purity. With lower purity the above values may be even higher. The higher the charging capacity of the adsorbent, the more accurate will be the model. If the charging capacity of the adsorbent is not fully used, or if the charging capacity is too small, a calculation of the suction capacity on the basis of the desorbed gas quantity or the gas quantity to be adsorbed, respectively, will have too optimistic results. The surface of the adsorbent quantity and the adsorber's size increase in the case of low charging or charging capacity, respectively. The suction capacity established according to the model has then to be corrected upwards with the increasing container volume. A further prerequisite for the application of the calculation model is, that the pressure/time-dependent course of evacuation over time is under concern.

Piping Pressure Loss Calculation

For the calculation of the losses resulting from inter-connected conduits and components one uses the conductance value L. The conductance value depends on the line diameter, the line length, the kind of flow and the desorption gas. For all pressure ranges in vacuum the calculation for round tubes and air is carried out according to Jaeckel:

$$L = \frac{3.6r3}{z} \ (2150 \ rp_m + 95) \ (m^3/h)$$

For laminar flows - which is the case if the lines are sufficiently long - the second term may be neglected.

$$L = \frac{7750 \ r^4 p_m}{z} \ (m^3/h) \tag{10}$$

where L = conductance value in m^3/h, S = required suction capacity at the container/adsorber, S_v = required suction capacity at the vacuum pump, p_m = mean pressure in mbar, r = tube radius in cm, z = tube length in cm.

For pipe bends and valves one uses their equivalent tube length.

The suction capacity of the vacuum pump is calculated on the basis of the conductance value L and the required suction capacity at the adsorber:

$$S_v = \frac{LS}{L-S}$$

With a relatively high desorption pressure, a mean p_m of e.g. 100 mbar, a relatively small suction capacity S and a compact design of an installation featuring small tube lengths and generously sized tube diameter, the conductance value will become relatively high, so that an application in the above equation will have as a result that S_v is approximately equal to S.

The vacuum conductance will be calculated for all vacuum pressure ranges in accordance to Jaeckel:

$$L = \frac{3.6r^3}{z} \ (0.039 \ \frac{rp_m}{\eta} + 30 \ \ \frac{T}{M}) \ (m^2/h)$$

where T = absolute temperature of the gas, M = molecule mass of the gas, and η = viscosity of the gas in Pas.

Example 1

Determination of the suction capacity of a vacuum pump for N_2 PSA installation for the production of 1000 m^3n/h nitrogen, 99.5%.

- Charge gas quantity
- Product gas quantity
- Off-gas quantity
- Temperature in evacuation
    condition
- Adsorption pressure
- Final desorption pressure
- Atmospheric pressure on site
- Pressure compensation pressure
- Adsorber volume
- Number of adsorbers
- Gap and clearance volumes
- Cycle tiem for adsorption and desorption 60 s each.

$V_2 = 3000$ m^3n/h air
$V_3 = 1000$ m^3n/h
$V_1 = V_2-V_3 = 2000$ m^3n/h

$T_1 = 293$ K
$P_a = 3$ bar abs.
$P_e = 90$ mbar
$P_o = 980$ mbar
$P_d = 1.5$ bar
9 m^3
2
$V_5 = 9m^3 \times 0.35 \times 60 = 189$ m^3n/h

Determination of V_o according to equation (4)

$$V_o = V_1 - V_5\left(\frac{1.5}{0.98} - 1\right) = 1900 \ m^3 NTP/h$$

Determination of V according to equation (5)

$$V = \frac{V_o \, P_n \, T_1}{P_o \, T_n} = 2107 \ m^3/h$$

$$V \triangleq 35.1 \ m^3/min \ and \ cycle$$

Determination of S_o according to equation (7) as o-approximation.

$$S_o = \frac{V}{t} \ln \frac{P_o}{P} = 1.465 \ m^3/s$$

Determination of the partial gas quantities according to equation (8), and course of evacuation and desorption of O_2 on carbon molecular sieve at S_o constant = 1.465 m/s.

1. Interval. $t_1 = 0$ S, $P_1 = 980$ mbar, $t_2 = 5$ S, $P_2 = 500$ mbar:

$$\overline{V} = \frac{S_o(t_1-t_2)(P_1-P_2)}{P_o \ln \frac{P_1}{P_2}} = 5.32 \ m^3$$

2. Interval. $t_1 = 5$ S, $P_1 = 500$ mbar, $t_2 = 11$ S, $P_2 = 300$ mbar:

$$\overline{V} = \frac{1.465 \times 6 \times 200}{980 \times 0.51} = 3.51 \ m^3$$

3. Interval. $t_1 = 11$ S, $P_1 = 300$ mbar, $t_2 = 18.5$ S, $P_2 = 200$ mbar:

$$\overline{V} = \frac{1.465 \times 7.5 \times 100}{980 \times 0.405} = 2.77 \ m^3$$

4. Interval. $t_1 = 18.5$ S, $P_1 = 200$ mbar, $t_2 = 26.5$ S, $P_2 = 150$ mbar:

$$\overline{V} = \frac{1.465 \times 8 \times 50}{980 \times 0.288} = 2.078 \ m^3$$

5. Interval. t_1 = 26.5 S, P_1 = 150 mbar, t_2 = 35 S, P_2 = 120 mbar:

$$\bar{V} = \frac{1.465 \times 8.5 \times 30}{980 \times 0.223} = 1.7 \text{ m}^3$$

6. Interval. t_1 = 35 S, P_1 = 120 mbar, t_2 - 50 S, P_2 = 100 mbar:

$$\bar{V} = \frac{1.465 \times 15 \times 20}{980 \times 0.1823} = 2.45 \text{ m}^3$$

7. Interval. t_1 = 50 S, P_1 = 100 mbar, t_2 = 60 S, P_2 = 90 mbar:

$$\bar{V} = \frac{1.465 \times 10 \times 10}{980 \times 0.1053} = 1.422 \text{ m}^3$$

Sum $\Sigma \bar{V}(1-7)$ = 19.25 m^3

The actual suction capacity is calculated according to equation (9).

$$S_s = S_o \frac{V}{\Sigma(\bar{V}_1 - n)} = 2.67 \text{ m}^3/\text{s}$$

Consequently, the required mean suction capacity S_s is 2.67×3600 = 9612 m^3/h.

Example 2

Determination of the suction capacity of a vacuum pump for O_2 PSA installation for the production of 1000 m^{3}n/h oxygen, 93%.

- Charge gas quantity V_2 = 12555 m^{3}n/h air
- Product gas quantity V_3 = 1000 m^{3}n/h
- Off-gas quantity V_1 = $V_2 - V_3$ = 11555 m^{3}n/h
- Temperature in evacuation
    condition T_1 293 K
- Adsorption pressure P_a = 1 bar abs.
- Ultimate desorption pressure P_e = 200 mbar
- Atmospheric pressure on site P_o = 980 mbar
- Pressure compensation pressure   1 bar abs.
- Adsorber volume   48.3 m^3
- Number of adsorbers   3
- Gap and clearance volumes   V = 48.3 m^3×0.35×60 = 1014 m^{3}n/h
- Cycle time for adsorption and desorption 60 s each.

V_o and V are determined using Eqs. (4) and (5), respectively, to be 11535 m^3/h (V_o) and 214 m^3/min (V).
    So is then calculated to be 5.67 m^3/s. The $\bar{V}$ values for the following 7 intervals are calculated: (1) t = 0–5s, P = 980–950 mbar; (2) t = 5–11 s, P = 850–700 mbar; (3) t = 11–16s, P = 700–600 mbar; (4) t = 16–23 s, P = 600–500 mbar; (5) t = 23–32.5s, P = 500–400 mbar; (6) t = 32.5–44 s, P = 400–300 mbar; (7) t = 44–60 s, P = 300–200 mbar. The sum of $\bar{V}$ = 144.5 m^3 and mean suction capacity S_s = 30240 m^3/h.

For examples 1 and 2 it has been assumed, that the piping
leading from the adsorber to the vacuum pump has been sufficiently
sized, so that the conductance value is high in relation to the
suction capacity, and that therefore the pipeline losses are low and
S_v is approximately equal to S_s. The values of the determination of
the suction capacity according to the above calculation model, or
according to examples 1 and 2, will be 5 to 10% higher than the
values determined in experiments. This means, that the calculation
model is always on the safe side.

After the suction capacity has been determined according to
example 1 or 2, one has to select a suitable vacuum pump, or define
the number of stages and the ratings of the individual pump stages.

Because of the required high suction capacity, only roots
vacuum boosters should be used as vacuum pumps, since they feature
up to ultimate pressure a quite constant suction capacity which
decreases only slightly. But these roots boosters are also very
advantageous with regard to power consumption, e.g., compared to
water ring pumps.

Function of the Roots Pump:

Vacuum technique uses roots pumps with double-teethed rotary pistons.
The basis functioning is shown in Fig. 4. Two 8-shaped pistons
rotate in opposite directions in a casing. Positive coupling via a
pair of gear wheels with identical teeth number causes an inter-
laced movement of the pistons which do not come into contact with
each other or with the casing wall. The resulting gaps between the
pistons, or between the wall and the pistons, are kept as small
as possible. They depend on the size of the pump, the desired
high efficiency, and the intended conditions of use. The actual
gaps are a compromise in the scale of approximately 0.1 mm.

In order to give a simplified representation of the pump's
function, fig. (4) shows only the right half of the device. When
the pistons are in positions I and II the pump volume facing the
container is increased. In position II the sickle shaped space V_2
is separated from the suction side. When the rotation continues,
the space facing the pressure side (pre-vacuum side) opens up and
the gas under pre-vacuum pressure P_v flows into the space which was
separated before (piston position IV). The incoming gas compresses
the gas contained in the space; when the piston rotation continues,
this gas is forced out together with the gas quantity which was
beforehand conveyed from the suction side. Therefore, if one
neglects the losses, the conveyed gas volume corresponds to the
volume V_2 of the sickle shaped space of position III. Owing to the
fact that this volume is built up twice with every rotation, and
considering furthermore that there are two pistons (in the left
part of the pump this volume V_2 also occurs twice per rotation), the
working volume (corresponding to the pump lift volume of reciprocat-
ing pumps) of roots pumps is:

$$V_S = 4V_2 n \qquad (11)$$

Power requirement:

$$W = \int P dV = \int_{P_a}^{P_v} V\, dP$$

Formula for the Rating of Rotary Piston Blowers

Flow of intake volume: $\quad S_1 = S_o - S_v\, m^3/min$ $\qquad\qquad$ (12)

Theoretical intake
volume flow: $\qquad\qquad S_o = n q_o/1000\ m^3/min$ $\qquad\qquad$ (13)

Loss volume flow: $\qquad\quad S_v = S_{v100}\ \dfrac{1.293 P}{\rho_1\ 100}\ m^3/min$ $\qquad$ (14)

Specific weight in
intake condition: $\qquad\quad \rho_1 = \rho_N P_1 T_N/(P_N T_1)\ \ kg/m^3$ $\qquad$ (15)

Temperature increase in
the blower: $\qquad\qquad \Delta t_{th} = \dfrac{\Delta p}{G_p \rho_1 n_v 10}\ \ °C$ $\qquad\qquad$ (16)

Working temperature: $\quad\ t_2 = t_1 + \Delta t_{th}\ \ °C$ $\qquad\qquad$ (17)

Volumetric
efficiency: $\qquad\qquad\quad n_v = S_1/S_o\quad \%$ $\qquad\qquad$ (18)

Theoretocal actuation
performance: $\qquad\quad W_{th} = S_o \Delta p/600\quad kW$ $\qquad\qquad$ (19)

Coupling performance: $\quad W = W_{th} + W_v\quad kW$ $\qquad\qquad$ (20)

where n = blower speed (1/min); q_o = blower volume (1/r); S_{v100} =
loss volume flow (at Δp = 100 mbar and ρ_1 = 1.293 kg/m^3) (m^3/min);
ρ_N = specific weight in standard condition (at 1.013 bar 0°C)(kg/Nm^3);
Δp = pressure increase (mbar); C_p = specific heat, (kJ/kg/°C); P_N =
1.013 bar (at 760 Torr, 0°C); W_v = loss performance.

Selection of Vacuum Pump System

One can select between so-called roots vacuum pumps with pre-intake
cooling or water-injected pumps. If the ultimate desorption pressure
is below 100 mbar and the product gas is at the same time the
desorption gas which has to be dry, one can use roots pumps with pre-
intake cooling. If the desorption gas can be wet, and if the ultimate desorption pressure is not far below 100 mbar, one uses the
water-injected roots pump, which with regard to power consumption, is
by far superior to the pump with pre-intake cooling.

If one compares the roots pump with pre-intake cooling with the
water-injected pump one can see that the volumetric efficiency of the
version with pre-intake cooling is worse than that of the water-
injected version. The reason therefore is, that the compression
temperatures of the water-injected version can be kept lower because
of the evaporative cooling, whereas the pump with pre-intake cooling
is only cooled by a recriculating cooler.

Owing to the higher temperatures in the pump chamber the gas density is reduced; therefore more gas flows through the gap back to the suction chamber. As a result the volumetric efficiency becomes worse. In the case of a roots pump with pre-intake cooling the power consumption is additionally increased by the fact that the cooling as flowing into the pump chamber has also to be conveyed in addition to the pumping gas.

Owing to the flow of cooling gas into the pumping chamber of the pump with pre-intake cooling, the compression operation is anticipated prior to the opening of the pump chamber towards the pressure pipe joint of the compression gas side. As a result the rotary pistons have to transport compressed gas at an earlier time, and this gas quantity is even increased by the quantity of the cooling gas.

Example 2 shows the rating of a vacuum pump station of a mean suction capacity of 30,000 m^3/h; initial pressure 980 mbar, ultimate pressure 200 mbar.

Calculation Steps Using the Formula for the Rating of Rotary Piston Blowers:

1. The corresponding mean pressure is determined for the mean suction capacity: P_e = 980/2 - 200 = 300 mbar.

2. Selection of the stages: If one compares two-stage pumps with single-stage versions, the clear benefits with regards to energy consumption manifest only from an ultimate pressure of approximately 300 mbar onwards. Three-stage pumps make only sense from an ultimate pressure of less than 100 mbar onwards. Therefore we select a two-stage pump station. With an ultimate pressure of 200 mbar, the stage ratio should not be smaller than 1:2. The ratio selected for the ultimate pressure under concern, i.e. 200 mbar, is 1:2.5.

3. Determination of the pump stages: Second stage: S_1 = 200 m^3/min at 500 mbar.

Type of pump according to manufacturer's documentation: Type 18.17, S_{v100} = 15 m^3/min; q_o = 365 l/rotation.

First stage: S_1 = 30000 m^3/h bei 200 mbar

$\cong$ 500 m^3/min

Pump type according to manufacturer's documentation:

Type 20.20, S_{y100} = 28 m^3/min; q_o = 1098 l/rotation (see Fig. 5).

4. The known operating point (1) of 30000 m^3/h is entered at 200 mbar in the double-logarithmic plot (Fig. 6). Under the condition that the mass flow is constant for the resp. operating point (Fig. 2, $Q = S_o p_o$ = constant) a straight line is drawn under an angle of 45°. With 12000 m^3/h operating point (2) is that of the second pump stage at 500 mbar.

5. Calculation of the required nominal suction capacity for the first and second stage:
First stage: $\rho_1 = \rho_n (P_1 T_N)/(P_N T_1)$ = 0.2378 kg/m^3.

Without water injection: $S_v = S_{v100}(0.01293\Delta P/\rho_1)^{1/2} = 113$ m^3/min.
$S_o = S_1 + S_v = 500 + 113 = 613$ m^3/min $= 36780$ m^3/h

The operating point (3) is entered at 1000 mbar in Fig. 6. Speed n = 1000 $S_o/q_o = 558$ min^{-1}.

Second stage: $\rho_1 = 0.594$ kg/m^3 and without water injection: $S_v = 49.5$ m^3/min, $S_o{}^1 \cong 14970$ m^3/h.

Operating point 4 is entered at 1000 mbar in Fig. 6. Speed n = 683 min^{-1}.

6. Determination of further operating points for the first and second stage.

Second stage: for Δp 300 mbar: $\rho_1 = 0.832$ kg/m^3, $S_v = 32.6$ m^3/min and $S_1 = 13014$ m^3/h.

Operating point 5 is entered in Fig. 6.

First stage: Under the condition, that mass flow Q = constant, a straight line is drawn under an angle of 45° from operating point 4 to determine operating point 6.
    At first, an estimation of ΔP of the first stage is made for this operating point. ΔP estimated at e.g. 450 mbar. $\rho_1 = 0.654$ kg/m^3; $S_v = 83.5$ m^3/min and $S_1 = 31770$ m^3/h. This point is not within the line Q = constant.
    New estimation of P = 520 mbar: $\rho_1 = 0.571$ kg/m^3; $S_v = 96$ m^3/min and $S_1 = 31000$ m^3/h.
    This point actually is on the line Q = constant, and this is the operating point 6 in Fig. 6. In the case of water injection, the mass flow changes, so that the second stage has to pump the occurring water vapor too. The angle of mass flow will then change from 45° to approximately 51°.

Differential Pressure and Energy Consumption: (Two Stages)

Theoretical suction capacity 1 stage: $S_o = 36.780$ m^3/h $\,\hat{=}\, 613$ m^3/min.

Theoretical suction capacity 2 stage: $S_o = 14.970$ m^3/h $\,\hat{=}\, 249.5$ m^3/min, and W = S_qP/600. The calculation procedure is shown on the table on the next page.

Differential Pressure and Energy Consumption: (For One Stage)

Theoretical suction capacity 1 stage: $S_o = 36.780$ m^3/h $\,\hat{=}\, 613$ m^3/min and W = S_oP/600.
    A similar calculation can be made for one-stage case, and the total power $W_{tot} = 605$ KW.

Time (s)   Pressure (mbar)

	P_1	P_1	P_2	P_2	W_1 kW	W_2 kW	W_v kW	W_{tot} kW
0.00	1000.00							
3.00	906.00	176	–	–	180	–	32	212
6.00	824.06							
9.00	751.52							
12.00	686.93	313	–	–	320	–	32	352
15.00	629.77							
18.00	579.34	421	–	–	430	–	32	462
21.00	534.41							
24.00	494.46	506	–	–	517	–	32	549
27.00	458.17							
30.00	424.63	496	920	80	507	33	32	572
33.00	393.65							
36.00	364.79	446	810	190	456	79	32	567
39.00	338.21							
42.00	313.50	387	700	300	395	125	32	552
45.00	290.75							
48.00	269.68	350	620	380	357	158	32	547
51.00	250.24							
54.00	232.28	328	560	440	335	183	32	550
57.00	215.58							
60.00	200.17	300	500	500	306	208	32	546

Total energy consumption: W_{tot} 491 kW

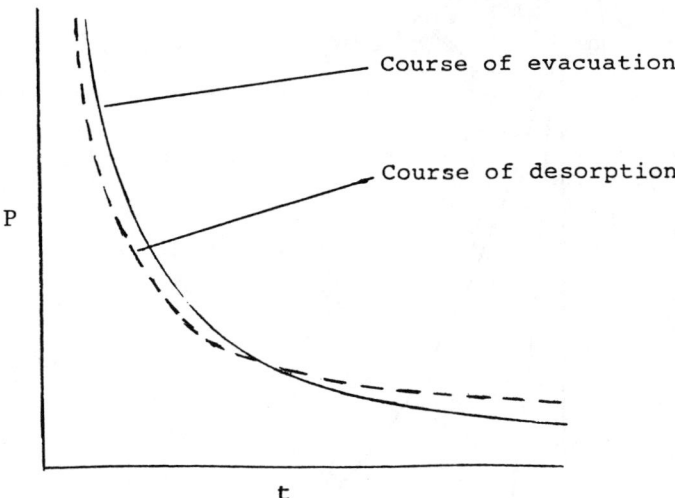

Figure 1. Course of evacuation vs. course of desorption.

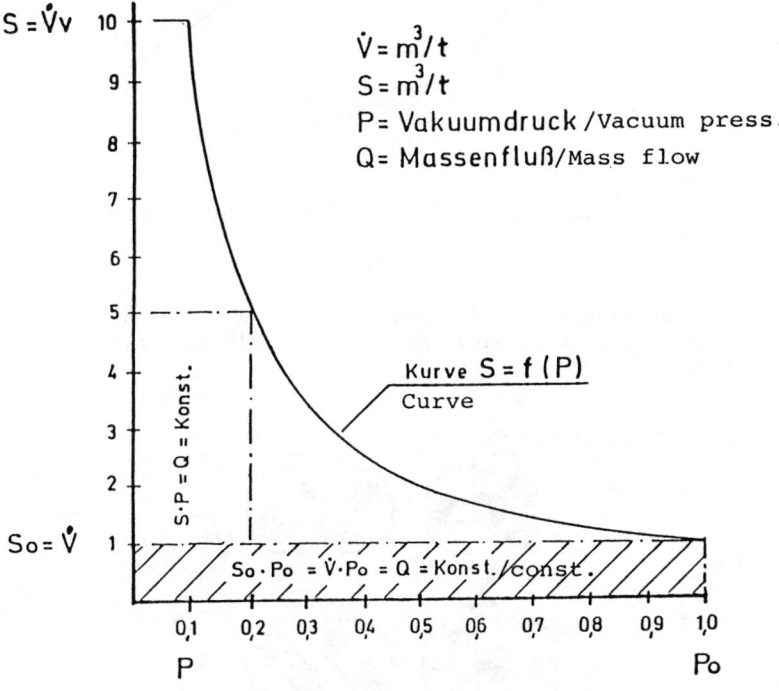

Figure 2. Representation of the change in suction capacity with the change in pressure. Note a log-log plot of S-P will result in a straight line.

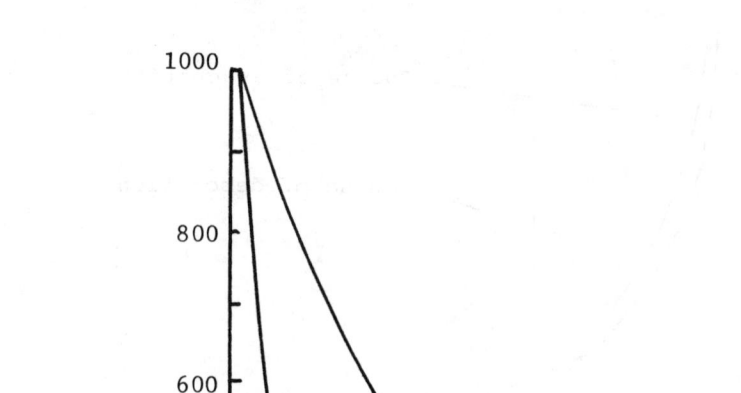

Figure 3.   Evacuation/desorption course of (A) N_2 on zeolite and (B) O_2 on carbon molecular sieve at approximately 20°C after limit charging.

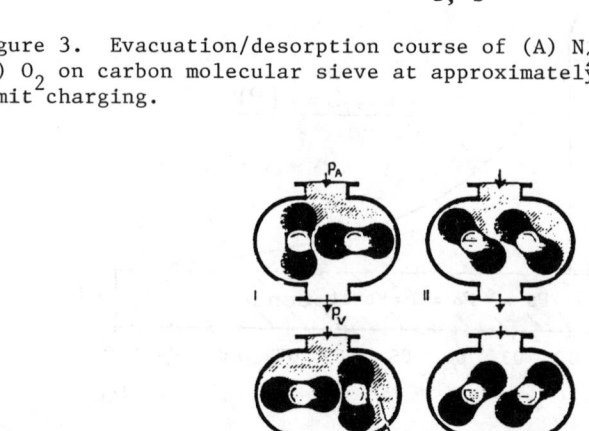

Figure 4.   Schematic of roots pumps.

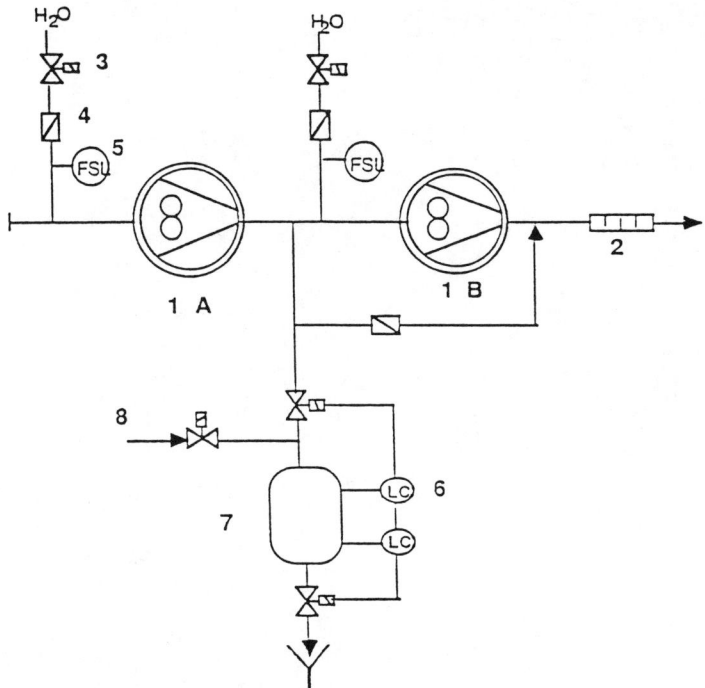

Figure 5. Pumping unit for oxygen generation plant. 1: water-injected rootspump, 2: adsorption silencer, 3: solenoid valve, 4: one-way valve, 5: flow switch, 6: level controller, 7: condensate tank and 8: airing valve.

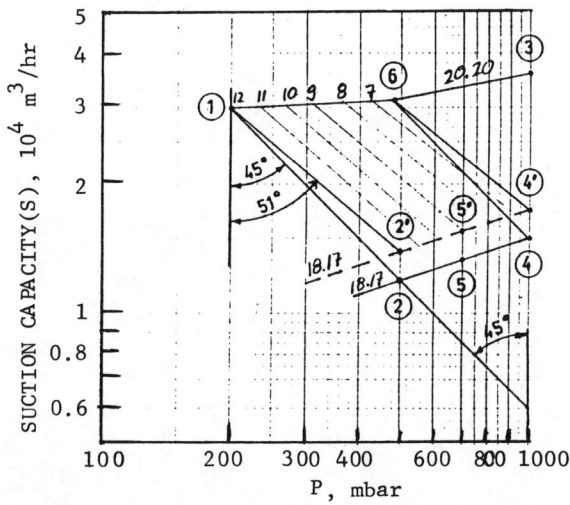

Figure 6. Calculation procedure for rating of rotary piston blowers.

Chapter 7

Evaluation of Macroreticular Resins as Gas/Vapour Sorbents to Rival Active Carbons

J. Hearn
P.L. Smelt
M.C. Wilkinson

Macroreticular resins have been tested for their ability to remove a range of gases and vapours under conditions simulating use in vapour filters. Typical charcoals were similarly evaluated for comparison purposes. Sorbents were characterised with respect to pore size distribution by nitrogen adsorption/desorption studies at $-196^{o}C$ and by mercury porosimetry. The performance of resins, which are predominantly mesoporous but have some microporosity, rivals that of charcoals, which are predominantly microporous with some mesopore contribution, only in wet (80% RH) air streams but is markedly inferior under dry conditions. For resins there was a general indication of increased performance with increasing surface area. For any given resin or charcoal there was a distinct decline in performance with decreasing sorbate boiling point.

INTRODUCTION

Carbons have historically dominated the field of respiratory protection in military and civilian use (1). However, there are polymeric alternatives which have surface areas approaching 1000 $m^2 g^{-1}$ which could potentially rival them; these include porous resins, functionalised resins and carbonised polymeric resins. Hydrophobic resins can be produced having permanent porosity in the dry state, ie. macroreticular resins, and these contrast with traditional ion exchange resins which possess a gel porosity only when solvated. Macronet resins (2) have a higher swelling capacity than macroreticular resins and are made by post reaction further crosslinking of swollen macroreticular resin beads. Functionalized resins can be produced from macroreticular resins by processes analogous to those used in ion exchange resin production, eg. harsh chemical treatment such as sulphonation using oleum. Surface functionality can have advantages in terms of specific adsorptive interactions or by surface catalytic reactions in noxious chemical treatment but this is at the cost of greater surface hydrophilicity which is disadvantageous in humid situations. Functionalized resins have also been carbonised with a view to carrying through desirable structural

111

features whilst exploiting the inherently superior adsorptive properties of the carbon surface (3). The disadvantages of carbons per se have been listed (4) as

(a)   high moisture uptake in humid atmospheres,
(b)   poor reproducibility in manufacture,
(c)   splitting and structural breakdown arising from fine pores in a rigid structure,
(d)   low steady state capacity,
(e)   inactivation by surface polymerisation of organic monomers,
(f)   inefficient cleaning for regeneration by steam,
(g)   slow adsorption kinetics.

In contrast, macroreticular resins have hydrophobic surfaces and in principle the pore characteristics can be tailor-made. Also the resin matrix will be capable of considerable swelling and regeneration by steam stripping, as used in monomer removal from polymers, is likely to be efficient. It should be possible to exert some influence on the kinetics of adsorption by control of the pore size distribution.

The swelling of the polymeric adsorbents is given by Miller (5) as

$$U = U_O + D$$

where U = equilibrium volume swelling per unit mass of dry polymer

U_O = U for the equivalent polymer made without diluent

D = volume of diluent

Thus for resins, vapour sorption is a combination of physical adsorption, pore filling and swelling of the cross linked matrix. Therefore even a non-porous resin has the potential for considerable vapour uptake. This potential will be fully realised in the static equilibrium situation but kinetic factors will determine its importance in the dynamic flow situation of filters where surface adsorption is likely, at least initially, to be more important.

DEVELOPMENT OF POROUS POLYMERS

In the 1950s and early 1960s many patents where filed (6) concerning the preparation, by suspension polymerisation, of crosslinked styrene divinyl benzene copolymers in the presence of diluents. This was mainly with a view to improving the performance of ion exchange resins or to mimic in non aqueous media the "gel filtration" or molecular sieving behaviour shown in aqueous media by crosslinked dextrans(7). Moore(8) preferred the term "gel permeation" for molecular separation by styrene divinyl benzene copolymers which has been widely adopted in modern GPC. A 1967 paper by Seidel et al (6) reviews, in German, the literature up to that date with some views on the development of the polymerisation reaction and the mechanism of formation of the porous morphology. Gels having dry porosity have been variously described as "macroporous" (5), "macroreticular" (9) and "visibly porous" (10). The last mentioned term, arising from the visible apperance of the beads (ie. beads having porosity are opaque white or transluscent depending on the pore size and extent of light scattering rather than any implication that the pores are visible to the naked eye) has not been widely adopted. Both "macroreticular" and "macroporous" are still widely used with the latter possibly finding more favour recently (11). "Macro" applies to the amount of porosity, originally cited by Miller as > 0.1 ml/g of cyclohexane uptake. The current authors use macroreticular in order to avoid the confusion which could otherwise arise when the IUPAC convention (12) for pore size classification is adopted to describe resin porosity; when macropore, mesopore and micropore have specific meanings in terms of pore diameter ie. macropores >50 nm, mesopores between 50 and 2 nm and micropores <2 nm. It would be better to avoid the confusion of macroporous resins which in fact contain no macropores and to avoid using microporosity to imply a low level of porosity in a resin which contains no permanent micropores in its dry state. If "macroreticular" is taken to imply a large scale network structure within the resin beads then we can find, in view of the polymer morphology discussed later, and because it is a well established term, no real objection to its use. There is some evidence (13,14) that the term which will find increasing favour will be simply "porous polymers".

As discussed in Seidel's review (6) macroreticular polymers are prepared by the copolymerisation of styrene and divinyl benzene in the presence of inert components (diluents) which can subsequently be removed from the end product. The diluent needs to be unreactive, so as not to be permanently bonded into the polymer structure, but must be miscible with the monomers, whilst having very low solubility in water so as not to be leached into the continuous (aqueous) phase during the reaction. A large number of organic chemicals fulfil these conditions. Basically, non solvating diluents produce large total pore volumes but low specific surface areas whilst diluents which are good solvents for polystyrene produce high surface areas and moderate total pore volumes. Polymeric diluents such as preformed polystyrenes of different molecular weights can also be used as porogenic agents, tending to produce rather low specific surface areas but offering a means of controlling mean pore size and tending to produce large pores. Not only does the concentration of diluent used control porosity but it is also influenced by the concentration of cross linking agent employed. When the amount of DVB used and diluent concentraiton are plotted on axes on a composition diagram (15), then various product regions can be identified ie. gel polymer, which shows no permanent dry porosity, at low DVB and low diluent content at one extreme and mechanically unstable powdery material at high diluent contents at the other extreme. In between, increasing levels of permanent porosity arise from increased diluent and increased crosslinker concentration.

The copolymerisation reaction is somewhat complex since commercial DVB is not only an isomeric mixture (p DVB and m DVB) but also a 55% volume solution in ethyl vinyl benzene (p and m isomers). P DVB is more reactive than m DVB but styrene and the ethyl styrene isomers are believed to have similar but lower reactivities (16). An observed enhanced reactivity of residual double bonds has been reviewed by Dusek (17) in terms of intramolecular cyclisation reactions whereas Guyot (14) reiterates Millar's view (5) that these are consumed in a further copolymerisation with styrene which swells the existing network to produce an interpenetrating polymer network.

In terms of the bead morphology produced, Kun & Kunin (18) first suggested three levels of substructure. Polymer chains first agglomerated to give nuclei which although not directly observed were postulated to be around 5-20 nm diameter and these later formed microspheres of 60-100 nm diameter. Sederal & De Jong (19) distinguished microporosity as the interstices between nuclei inside the microspheres, mesoporosity between microspheres and macropores within the beads. Guyot et al (14) report that when the porogenic agent is a solvent then they identify only large nuclei (20-50 nm diameter) and agglomerates but with non solvent agents the three levels of sub structure described by Kun & Kunin were evident. The degree of coalescence of the nuclei they identified as depending upon the timing of three critical events ie. gelation, overlapping (interpenetration of polymer chains in solution)and phase separation. The gel fraction first appears when the gel is swollen to its maximum extent and further crosslinking leads to syneresis. The concentration at which overlapping of chains occurs will increase as the crosslink density increases and the solvent quality of the medium decreases (monomer consumption). Precipitation also depends on solvent quality and crosslink density. Guyot explains the structure of high surface area resins prepared using a high diluent concentration of good solvent and at high DVB content as arising from gelation taking place before overlapping (with precipitaion not arising in the prevailing solvency conditions). The DVB is mostly consumed in the growth of nuclei which are sufficiently crosslinked to maintain their own identity when overlapping does occur. The nuclei, in the latter stages of growth, may be linked by linear polymer formed by styrene or ethyl styrene reacting with residual double bonds on the surface of the nuclei. Upon drying, considerable stress and shrinkage results in very small pores being formed between the nuclei. The above structure helps to explain the ability of these resins (solvent porogen) to swell considerably by expansion between nuclei in both solvents and non solvents and surprisingly more so in the latter.

Isoporous or macronet resins (20) are prepared by Friedel-Crafts alklyation to further crosslink lightly crosslinked macroreticular resins and these display similar properties, ie. high surface areas with a very narrow distribution of fine pores, as macroreticular resins produced in the prescence of solvent. In this case, however, the pore volume and surface area are very dependent on the last treatment the macronet resin received (14) with the highly stressed structure prone to collapse upon drying following swelling.

CHARACTERISATION OF PORES

Following a suitable procedure to remove residual monomers and diluents from the resin beads, such as Soxhlet extraction with methanol followed by drying, then the techniques which have been commonly used for pore characterisation are nitrogen adsorption/desorption at 77K and mercury penetration over a range of applied pressures. Nitrogen adsorption data can be used to calculate BET surface areas and the pore size distribution. In the mesopore range pore sizes can be obtained from the desorption branch of the hysteresis loop via the Kelvin equation using the BJH (21) or Dollimore and Heal (22) procedures to allow for adsorbed layer thickness and typically a cylindrical pore model. In mercury porosimetry pore size analysis is obtained via the Washburn equation after correcting the data for the compressibility of mercury (23). The pore radius ranges covered, and in both cases it is the size of the entrance to pores which is considered even though the body of the pore may be much wider, are between 2 nm and 50 nm for nitrogen desorption and between 3.7 nm at 2000 atmospheres pressure and 7260 nm at 1 atmosphere for mercury penetration. Some reservations have been expressed (13,14,24) about the structural damage caused to resin beads at the high pressures available in the latter technique.

The total pore volume can be evaluated from the mercury displacement density (apparent density) plus the helium displacement density (skeletal density) using the relationship:

$$\text{Total Pore volume} = (1/\rho_{apparent} - 1/\rho_{skeletal})$$

The Gurvitsch rule (25) states that the liquid volume of different adsorbates, when measured on porous adsorbents, is essentially constant and represents the filling of the total pore volume ie. if a relative pressure close to unity is used above the point at which the hysteresis loop closes then the total pore volume can also be calculated from the uptake of nitrogen converted to a liquid volume.

Micropore volume, more usually evaluated in charcoals, can be obtained by plotting nitrogen uptake against statistical layer thickness, derived from a number of sources eg. the De Boer equation or standard plots based on porous silicas and aluminas, and extrapolation to zero layer thickness. The intercept on the uptake axis is taken to be the total micropore volume ie. the "t plot method" (26). The α_s method is a similar technique where α_s replaces t and is set equal to unity at a p/p_0 of 0.4 (12).

The Dubinin-Radushkevitch equation:

$$\log W = \log W_0 - D \; \log (p/p_0)^2$$

where W = amount adsorbed as a liquid volume
W_0 = total micropore volume
D = a characteristic constant at a given T

produces linear plots of $\log W$ vs $\log (p/p_0)^2$ which have been extrapolated, for microporous carbons, to yield total micropore volumes (12).

Gregg & Langford (27) used the pre adsorption of nonane to block micropores but leave mesopores and external surface clean, when, the difference between adsorption isotherms in the prescence and absence of nonane yields the micropore volume.

APPLICATIONS

Considerable current interest is shown (28 & 29) in porous styrene DVB polymers for use as supports in fine chemical synthesis, following Merrifield's (30) complex oligopeptide sequential synthesis. When used as supports for chemical reagents or in catalysis it is the ease of recovery, by filtration, of the supported material which is the attraction.

Very little has been published in the open literature, although considerable interest is known to exist (31), on the use of macroreticular resins for air filtration and sorption of noxious chemicals.

The current study has been limited to macroreticular resins only and in the main to styrene-divinylbenzene copolymers, some of which are commercially available, and others which were prepared in our laboratories. These were compared with a range of typical charcoals for their ability to remove a range of gases and vapours from dynamic air streams at 0% and 80% RH.

EXPERIMENTAL

MATERIALS

Styrene, divinylbenzene (DVB) and azobisisobutyronitrile (ABIN) were all laboratory grade (BDH Ltd, UK). The styrene was distilled at reduced pressure and ABIN was recrystallised from methanol, DVB (52% DVB (m and p), 46% EVB, 2% DEB) was washed with sodium hydroxide solution to remove inhibitor. Methyl cellulose (Polysciences, USA) and polyvinylpyrollidone (BHD Ltd, UK) were used as received. Commercial resins employed were XAD's (Rohm & Haas, USA) and Chromosorb Century Series (Johns Manville, USA).

RESIN PREPARATION

Resins were prepared in cylindrical baffled reaction vessels using a paddle type stirrer at 250 rpm.

CLEANING

Resins were cleaned prior to use by Soxhlet extraction with methanol for 8 hours.

CHARACTERISATION

Where possible, appropriate sieve fractions were obtained using Endecott sieves. Pore radii distributions were obtained between 50 μm and 37A° using Macropore and Porosimeter 2000 units (Carlo

Erba, Italy). Nitrogen adsorption at 77K was performed using either, for one point surface area determinations a Strohlein Area Meter or for full adsorption/desorption isotherms, a conventional BET apparatus. Pore size analysis data was carried out by MCA Services, Cambridge, England. Static adsortion data was obtained for saturated halocarbon vapours by removing samples for weighing from a thermostatted dessicator. Dynamic adsorption measurements were made using the apparatus previously illustrated (24). The resin or carbon samples were contained in 'volume activity' tubes, ie. brass cylinders of 2 cm diameter and lengths in the range 0.5 to 5 cm which were filled using a 'snowstorm filler' - a funnel device which utilizes cross wires to set particles into a circulatory motion as they fall so that they pack uniformly. The samples, preconditioned at 0% and 80% RH could be exposed to an air flow of 1 dm^3 min^{-1} into which could be introduced 2 mg/dm^3 of halocarbon gas or vapour from a hot finger device. The air stream could be maintained at 0% or 80% RH and was monitored using a Panametrics RH2T detector (Panametrics, UK) whose performance was checked by adsorption of water vapour onto P_2O_5. The effluent halocarbon was monitored by using initially a CDE Mk4 Halogen Detector and latterly a VG Instruments (UK) mass spcetrometer.

RESULTS AND DISCUSSION

Tables 1, 2 and 3 show data on a number of resins and charcoals showing the penetration time $\frac{C}{C_0} = \frac{1}{100}$ for carbon tetrachloride vapour, the surface area, the total pore volume from Hg penetration, static CCl_4 uptake at 0% RH and water vapour uptake at 80% RH. The conclusions which can be drawn are as follows:

(a) at 80% RH the penetration time for resins approach those for carbons under the same conditions.
(b) charcoals absorb much more water at 80% RH than the resins,
(c) the performance of the in-house resins approach but are slightly inferior to the commercial resins,
(d) the static uptakes for resins even at 0% RH exceed those of charcoals in contrast to the dynamic behaviour.

Table I   Trent Resins

SAMPLE Code-pls	PEN.TIME CCl_4 2mg/1 80% RH (mins)	SURFACE AREA (m^2/g)	Hg PORE VOLUME (mm^3/g)	CCl_4 UPTAKE wt% 0%RH	H_2O UPTAKE wt% 80%RH
2	13.5	–	346	58.4	0.21
3	23	–	336	70	0.64
4	8.5	–	305	65	0.36
5	13.3	–	378	62.5	1.02
6	13.3	–	441	34.4	0.16
8	27	392	1800	33.8	0.19
9	35.5	326	840	32.2	0.80
10	33	527	640	14.7	0.36
11	32	307	1351	18.3	0.04
12	18.5	258	1363	15.6	0.45
17	10	318	358	47.5	1.83
18	30	355	269	55.5	1.46
19	3	103	137	29.5	0.81
20	12	419	555	53.6	0.62
21	2.5	5	160	58.9	5.20
22	2.5	5	49	17.6	0.19
23	2.5	5	57	1.1	–
24	2.5	5	134	52.7	1.97

Table II   Commercial Resins

SAMPLE	PEN.TIME CCl_4 2mg/1 80% RH (mins)	SURFACE AREA (m^2/g)	Hg PORE VOLUME (mm^3/g)	CCl_4 UPTAKE wt% 0%RH	H_2O UPTAKE wt% 80%RH
C–102	16.5	350	848	27.8	0
C–104	13.3	150	207	15.4	1.53
C–105	17.5	170	242	27.1	0.91
C–106	58	650	938	99.1	0.41
C–107	32.3	450	1502	43.8	4.55
XAD–2	22.3	330	637	30	5.04
XAD–4	36.8	750	758	100	1.62

Table III Charcoals

SAMPLE	PEN.TIME CCl$_4$ 2mg/1 80% RH (mins)	SURFACE AREA (m^2/g)	Hg PORE VOLUME (mm^3/g)	CCl$_4$ UPTAKE wt% 0%RH	H$_2$O UPTAKE wt% 80%RH
207 A	102.5	1050	366	30.4	11.2
207 C	94	1000	207	24.0	10.0
LS769	40.8	970	267	33.5	16.8
ASC12x30	42.0		211	36.8	9.5
NORIT	39		346	61.4	33.9
FPICA1367	27.5		229	49.6	32.6
PICA 135	33.8		234	56.5	33.8
CHEM ScII	14.3		185	50.1	15.7

Figure 1 shows that there is an indication that resin performance improves with increasing surface area, a trend which is better emphasised for samples at constant mesh size Figure 2.

There is no simple correlation between performance and total pore volume. Its effect is more likely to be apparent if samples were available at near constant surface area and variable pore volume. Static uptakes of water vapour show (32) type III isotherms for resins being typical of weak solid/gas interactions and type V isotherms for charcoals being associated with an increased micropore contribution. Undoubtedly it is the high water vapour uptake in the competitive adsorption situation which leads to the much reduced performance of charcoals when the micropore volume is reduced (having been filled with water). A more restricted range of samples were examined against a range of boiling point halocarbon challenges listed in Table 4. Linear relationships were found between performance and boiling point at both 0% and 80% RH (Figures 3 and 4) for both resins an charcoals. A similar linear correlation over a very wide range of boiling points has previously been reported for carbons (33). The results here further emphasise that at 0% RH charcoals have far superior performance. At 80% RH however the greater comparability of resins and charcoals and even the possible superior perfomance by a resin is shown.

Nitrogen adsorption data on charcoals and resins confirm a type I isotherm for charcoals (Figure 5) which is indicative of a high level of microporosity and a type IV isotherm (Figure 6) for resins, with a hysteresis loop signifying considerable mesoporosity. The failure of the loops to close at low $\frac{P}{P_0}$ on desorption is probably indicative of a failure to obtain true equilibrium on the adsorption cycle rather than a failure to remove nitrogen on desorption. Behaviour of this type is found in coals (34) being often indicative of ultra-microporosity.

Table V

Resin	Surface Area $(m^2 g^{-1})$	Volume $(cm^3 g^{-1})$	Mercury Pore Volume $(cm^3 g^{-1})$	Gurvitsch Pore Volume $(cm^3 g^{-1})$	Apparent/ Sketetal Pore Volume $(cm^3 g^{-1})$
XAD4	806	0.713		0.926	0.708

Similar isotherms have been reported previously for porous resins and other explanations presented (35) for the failure of the hysteresis loop to close. Howard cites Hoburg el al (36) who attributed the low pressure hysteresis of nitrogen in linear polystyrene to penetration of nitrogen into the molecular structure and suggests also that upon sorption of nitrogen to high relative pressures then the polymer structure opens somewhat only to collapse upon pressure reduction to yield activated diffusion effects upon desorption.

Total pore volume can be estimated by three approaches Table V (i) Hg penetration down to 37A° radius, (ii) He and Hg apparent densities below 3μm radius, and (iii) from the Gurvitsch volume up to 100 nm radius. It is interesting to note that whilst techniques (i) and (ii) yield comparable results, the Gurvitsch volume is higher. Similar results have been found for other resins (24). This could arise from solubility of nitrogen in the polymer or structural damage at 77K causing extra porosity.

Comparison of the data for the resin XAD 4 between N_2 desorption Figure 7 and Hg porosimetry Figure 8 shows poor agreement in the region of expected overlap, ie. between 100–10 nm diameter; nitrogen shows pores mostly below 10 nm but mercury penetration finds 0.55 ml gm^{-1} of pores between 100–10 nm(cf. 0.05 ml g^{-1} for nitrogen). The overall total pore volumes are similar but displaced in size. Possible explanations here are that forceful mercury intrusion stretches some pores with commensurate loss in volume from adjacent pores or that without penetration pore volume is 'squashed' out of the structure to give a useful estimate of total pore volume if not of individual size. That reasonable agreement in the region of overlap between the two techniques is reported for other materials (23) suggests that structural damage is the problem here.

DR plots were not amenable to analysis for total micropore volume but when t and α_s plots were applied to the commercial resin XAD 4 then the micropore volume was in good agreement with that deduced from the nonane pre adsorption technique (32). This type of analysis had not previously been reported in the literature for porous polymers.

Although the absolute significance of pore size data is questionable in view of the many approximations involved, it is the systematic differences between samples which are of interest in terms of screening materials of best dynamic performance and in trying to identify the critical morphological features. In carbons, capacity resides in the micropores with larger transport pores required for good dynamic performance which is optimized in charcoal cloth. For resins there is a trend of improved performance with increased BET surface area which does not however correlate with micropore volume (24) and it is anticipated that further work with model resins ie. of fixed surface area and variable pore volume distribution and of fixed total pore volume and variable surface area, will be needed in order to identify the critical factors in this case, in relation to dynamic (adsorptive) performance at relatively low partial pressures.

CONCLUSIONS

It is apparent that whilst currently available macroreticular resins show some promise in the field of vapour sorption in humid atmospheres and particularly in situations, irrespective of humidity, where equilibrium swelling can be achieved (provided the container is not ruptured by the increased volume), further development would be required for resins to compete with charcoals in the low bed depth dynamic applications where they are currently clearly inferior. Greater control of porosity is needed, especially at high surface areas in order to optimize performance and its correlation with pore structure. Conventional pore analysis techniques are seldom unequivocal and this is particularly true for resins.

LITERATURE CITED

1. Hassler, J. W. Activated Carbon; Chemical and Process Engineering Series; Leonard Hill: London, 1967; p3.
2. Barar, D. G.; Staller, K. P.; Peppas, N. A., J. Polym. Sci. Polym. Chem. Ed. 1983, 19, 1013-1024.
3. Ehigiamusoe, R. E.; Howard G. J. J. Appl. Polym. Sci. 1967 19, 3327-3340.
4. Kennedy, D. C. U. S. Patent 3 789 876, 1974.
5. Millar, J. R.; Smith, D. G.; Marr, W. E.; Kressman, T.R.E. J. Chem. Soc. 1963, 218.
6. Seidel, J.; Malinsky, J.; Dusek K.; Adv. Polym. Sci. 1967, 5, 113-213.
7. Flodin, P. Dextran Gels & Their Applications in Gel Filtration; Mei jels Bokindustri; Halmstad 1962.
8. Moore, J. C., J. Polym. Sci. Pt. A. 1964, 2, 835.
9. Kunin, R.; Meitzner, E., JACS. 1962, 84, 305
10. Seidel, J.; Malinsky, J.; Dusek, K., Plasticheskie Massey. 1963, 12, 7.
11. Referee's comment on paper submitted to JCIS
12. Gregg, S. J., Sing, K. S. W. Adsorption Surface Area and Porosity, Second Edition, Academic Press 1982.
13. Albright, R. L., Reactive Polymers. 1986, 4, 155-174.
14. Guyot, A.; Bartholin, M., Prog. Polym. Sci. 1982, 8, 277-332.
15. Haupke, K.; Pientka, J., Chromatog. 1974, 102, 117.
16. Schwachula, G., J Polym. Sci. 1975, C53, 107.
17. Dusek, K.; Galina, H.; Mikes, J., Polymer Bulletin. 1980, 3, 19.
18 Kun, K. A.; Kunin, R., J. Polym. Sci. 1968, A1(6), 2689.

19. Sederel, W. L.; De Jong, G. J.; J. Appl. Polym, Sci. 1973, 17, 2835.
20. Davankov, V. A.; Tsyurupa, M., Angew. Makromol. Chem. 1980, 91, 127.
21 Barrett, E. P.; Joyner, J. G.; Halenda, P. P. JACS. 1951, 73, 373.
22. Dollimore, D.; Heal, G. R., JCIS. 1973, 42, 233.
23. Lowell, S.; Shields J. E., Powder Surface Area and Porosity Second Edition; Chapman Hall: 1984; Chapter 11.
24. Hearn J.; Smelt, P. L.; Wilkinson, M. C. Paper to be presented at PSA 88 Analytical Division RSC. UK. Surrey University 19 & 20 April 1988.
25. Gurvitsch, L., J. Phys. Chem. Soc Russ. 1915, 47, 805.
26. Gregg, S. J.; Sing, K. S. W., Adsorption Surface Area and Porosity, First Edition, 1967.
27. Gregg, S. J.; Langford, J. F., Trans. Farad. Soc. 1969 65, 1394.
28. Hodge, P.; Sherrington, D. C. Polymer Supported Reactions in Organic Synthesis. Wiley; New York, 1980.
29. Mathur, N. K.; Narang, C. K.; Williams, R. E., Polymers as Aids in Organic Chemistry. Academic Press: New York, 1980.
30. Merrifield, R. B., JACS. 1963, 85, 3045.
31. Rohm & Haas Company Private Communication.
32. Hearn J,; Smelt, P. L.; Wilkinson, M. C. Paper accepted for publication in JCIS.
33. Nelson, G. O.; Harder, C. A., Amer. Ind. Hygiene Ass. J. 1974, 291.
34. Jones. L. F., MCA Services Cambridge UK Personal Communication.
35. Howard, G. J.; Midgley, C. A., J. Appl. Polym. Sci. 1981, 26, 3845.
36. Hoburg, R. F.; Handler, G. S.; Scholtz, J. J., JCIS. 1968, 27, 642.

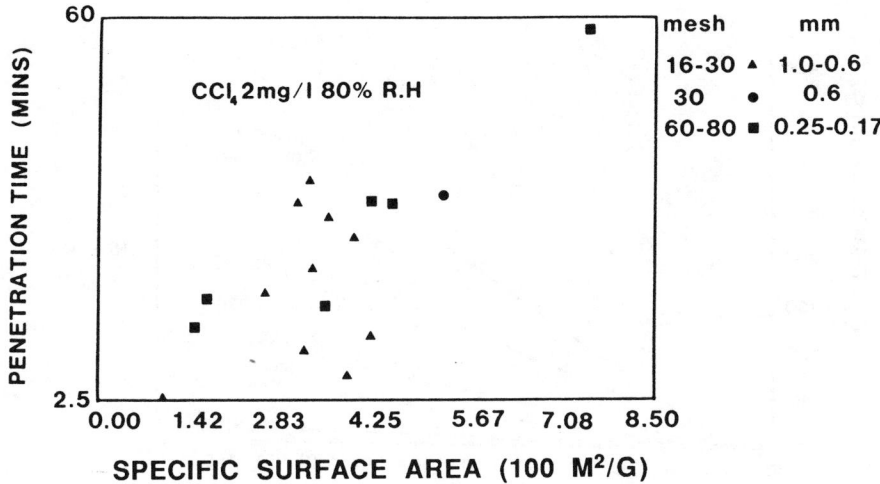

Figure 1   Penetration Time to CCl_4 (2 mg/l at 80% RH) for Resins
of   Variable   Surface Area (non constant mesh size).

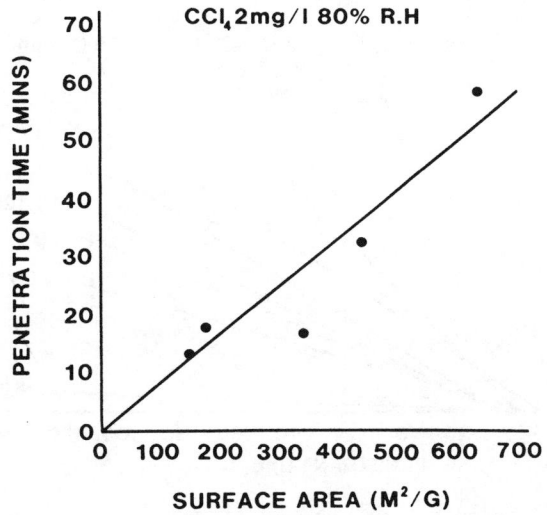

Figure 2   Penetration Time to CCl_4 (2 mg/l at 80% RH) for the
Chromosorb series of Resins (constant mesh size BSS
60-80).

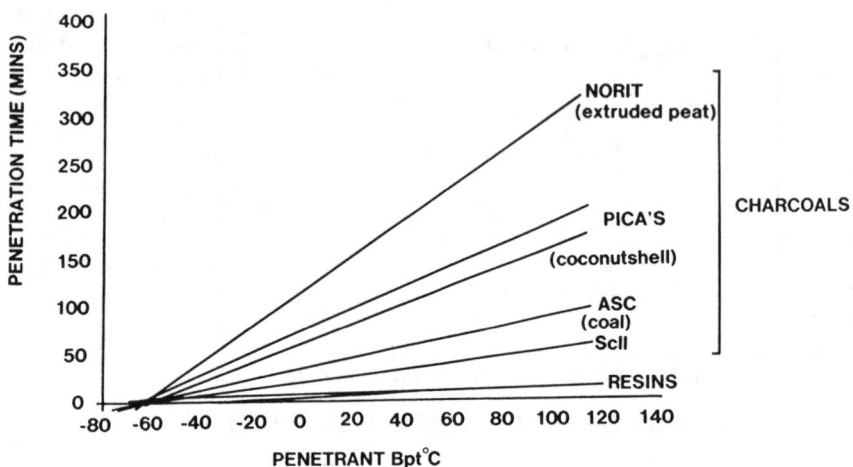

Figure 3   The Dependence of Penetration Time on Penetrant Boiling Point (0% RH)

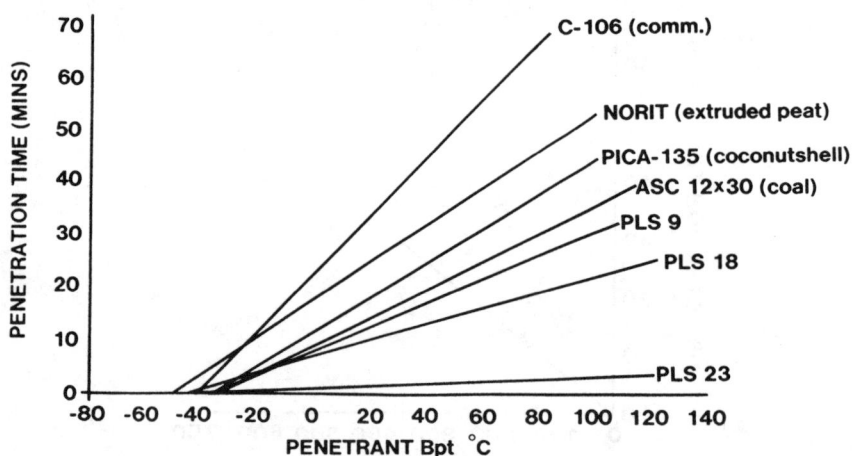

Figure 4   The Dependence of Penetration Time on Penetrant Boiling Point (80% RH).

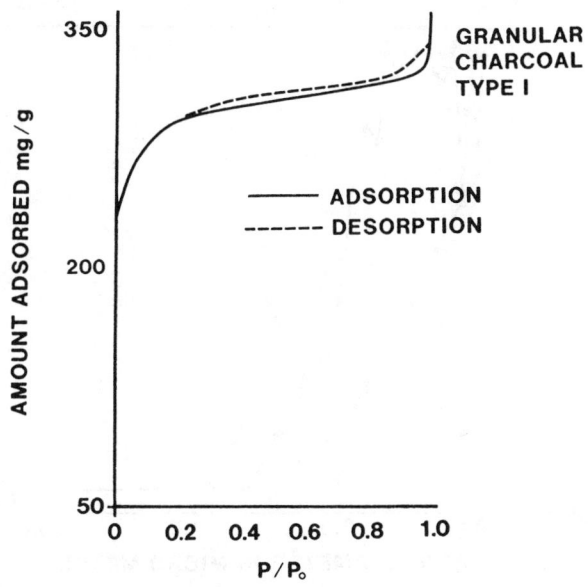

Figure 5   Nitrogen Adsorption–Desorption Isotherm for a Typical Charcoal.

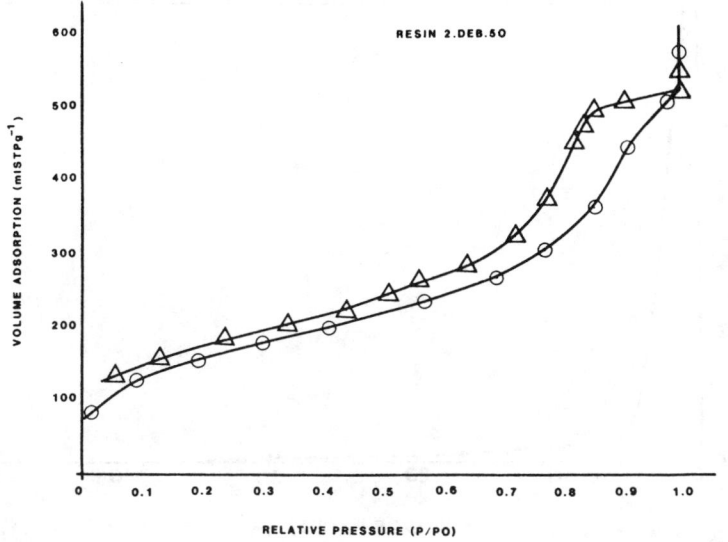

Figure 6   Nitrogen Adsorption–Desorption Isotherm for a Typical Resin.

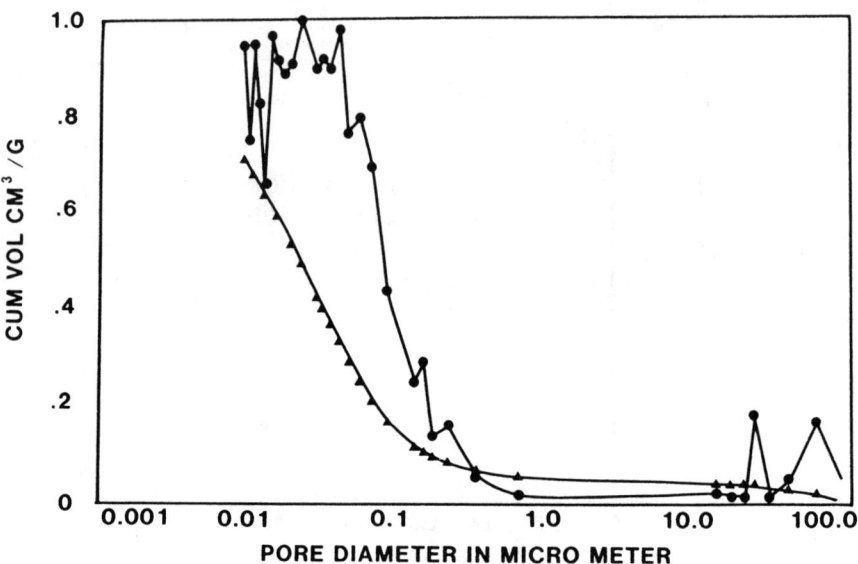

Figure 7   Cumulative and Differential Pore Size Distribution for Resin XAD 4 by Mercury Porosimetry.

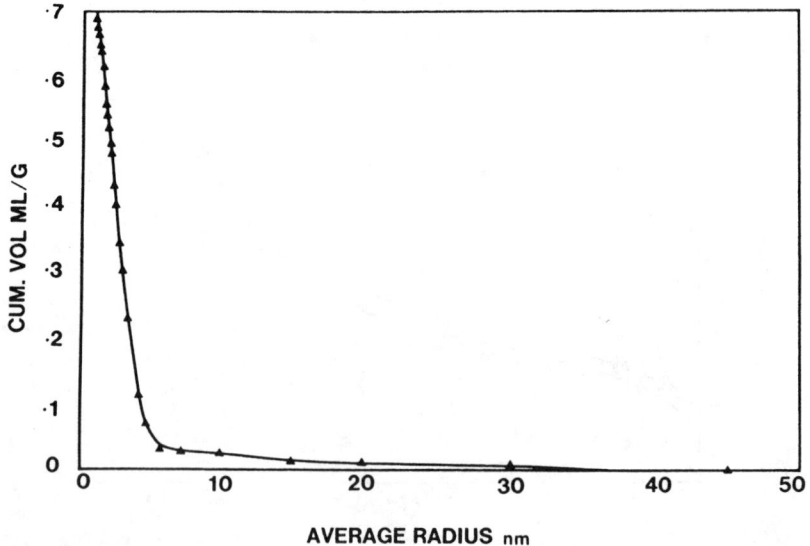

Figure 8   Pore Size Distribution for Resin XAD 4 by Nitrogen Desorption.

Chapter 8

Use of Surfactant-Enhanced Carbon Regeneration to Remove Volatile Organics from Spent Activated Carbon

Bruce L. Roberts
John F. Scamehorn
Jeffrey H. Harwell

Surfactant-enhanced carbon regeneration uses a concentrated surfactant solution to regenerate spent activated carbon. A water flush is then used to remove residual surfactant from the carbon bed. In this study, regeneration of carbon containing either toluene or amyl acetate was investigated. The process was found to effectively regenerate carbon containing either solute at various loading levels. Studies of the effect of regenerant solution flow rate and surfactant concentration indicate that the removal of toluene was nearly equilibrium limited, while amyl acetate exhibited significant mass transfer resistance. The residual surfactant on the carbon following the regeneration was readily removed by a water flush. The carbon showed no signs of serious degradation over nine regeneration cycles.

Introduction

Activated carbon is used to remove organics from water or gases in numerous applications (1-4). Carbon adsorber beds are widely used in clean-up of streams before emission to the environment and in recovery of organic products.

While carbon adsorption works efficiently at removal of many organics from effluent streams, the whole process is limited by regeneration of the carbon. Regeneration of activated carbon is a major factor in the cost effectiveness of the use of carbon (4). The standard method of regeneration, thermal regeneration (2), involves removal of the carbon from the bed followed by transport to a hearth regeneration furnace, where the adsorbed organics are volatilized and carbonized. This process is energy intensive, labor intensive, and time consuming. Further, the organic adsorbate is not recovered and a large fraction of the carbon may be burned in the furnace. An efficient, in-situ, regeneration method would be a great improvement over this standard regeneration method.

Hot gas regeneration is an in-situ regeneration method in which hot gas (e.g., steam or nitrogen) is passed through the bed to desorb the adsorbate by a combination of purging, and of desorption by heat-up effects (1,5). This is only effective when the adsorbate is highly volatile. Another in-situ regeneration method is solvent regeneration (6-9), in which an organic liquid solvent is passed through the bed to desorb the adsorbate. A major disadvantage of this is that when the process is complete, a hot gas regeneration must be performed to desorb the residual volatile solvent, making for an energy intensive process.

In biological regeneration (10), another in-situ regeneration method, bacteria are introduced into the bed to consume the adsorbed organic. Disadvantages include the process being very slow, the organic not being recovered, reduction of bed capacity from adsorption of some of the products of the degradation, the need to induce desorption of the bacteria when done, and, finally, the fact that the bacteria often cannot ingest a mixture of organics.

Surfactant-enhanced carbon regeneration (11) (SECR) uses surfactants (detergents) to remove adsorbed organics from activated carbon in order to regenerate it for reuse. In SECR, a concentrated surfactant solution is passed through the spent carbon bed. The adsorbate desorbs and is solubilized into micelles in the solution. Micelles are surfactant aggregates typically composed of 50 to 150 surfactant molecules. The micelle has a hydrocarbon-like interior into which organic molecules will dissolve or solubilize. A concentrated surfactant solution can contain large concentrations of dissolved organics through solubilization. Therefore, a small volume of the concentrated surfactant solution can potentially solubilize all of the adsorbed organic, resulting in a small stream which is highly concentrated in the organic adsorbate or solute. When the desorption process is complete, the carbon can contain some residual adsorbed surfactant. This is removed by a water flush. After the water flush is completed and all of the residual surfactant is removed from the carbon, the bed can be used directly for liquid phase applications or dried before reuse in vapor phase applications. The SECR process is illustrated in Figure 1.

Experimental

Methods. A 1" diameter, 3' long jacketed column with a filter at both ends was used for these experiments. The column contained 200 g of carbon. One pore volume in this column was 300 ml. The temperature was maintained at 30^0C with water circulated through the jacket from a heater-circulater in a constant temperature water bath. A plunger in the column adjusted the carbon bed depth and maintained a fixed bed height. A constant pressure gear pump maintained a constant flow rate as the regenerant fluid or flushing water was pumped through the bed.

The amyl acetate concentration in the effluent from the column was analyzed by gas chromatograph and the toluene concentration was analyzed using HPLC with a UV detector. The SDS concentration was analyzed using HPLC with a conductivity detector.

The organics were equilibrated with the carbon by bubbling compressed air through either toluene or amyl acetate at room temperature. The air/vapor mixture passed through the carbon bed and then to the condenser (cold finger) maintained at -20^0C, to detect if breakthrough occurred. The air was recycled to the compressor so that all of the vaporized organic was adsorbed on the carbon, making the adsorption level a known quantity. No organic condensation on the cold finger was observed during these operations. The amount of organic adsorbed on the carbon bed ranged from 5 ml to 20 ml. The equilibration time for the toluene was approximately 12 hours, and, for the amyl acetate, was approximately 24 hours.

If the regenerant solution was applied directly to the dry carbon, severe foaming was observed. This led to severe channeling or plugging of the column, due to air trapped between the granulated carbon particles. Air was also trapped in the pores of the carbon. Therefore, the column was flushed with water, before the surfactant solution was introduced, until air voids between the granulated carbon particles were no longer visually observed. The concentration of the desorbing solute in the effluent water was measured and used in calculation of the fractional removal of the solute.

During regeneration, the surfactant solution, containing between 0.1 M SDS and 0.30 M SDS, was pumped through the column at a flow rate of 2 mL/min to 15 mL/min. Analysis of the effluent and a knowledge of the amount of solute originally loaded on the column permitted a calculation of fractional recovery at any point in the run.

A water flush followed the regeneration step, to remove any residual surfactant from the carbon bed. In the flush, 20 L of water were pumped through the bed at either 10 mL/min or 20 mL/min.

Following the water flush, the bed was drained and the jacket temperature on the column was raised to 50°C. Compressed air was circulated through the bed and then passed through a condenser (cold finger) to condense the effluent water vapor. The air from the condenser was recycled back to the compressor. The drying was considered complete when condensate ceased to appear in the condenser. The drying step took from 12 to 24 hours.

To avoid start-up effects due to using virgin carbon (11), three complete cycles of adsorption and regeneration were completed before quantitative experiments were conducted. In these pretreatment runs, 20 mL of organic were loaded on the carbon, then 14 L of 0.2 M SDS regenerant solution and 20 L of water flushing solution were run through the column. The normal drying procedure was used.

Materials. The surfactant was sodium dodecyl sulfate (SDS) from Aldrich. The toluene was Reagent grade from Fisher. The amyl acetate was "purified" grade from Mallinckrodt. The activated carbon was a standard vapor phase carbon, PCB 12X30 from Calgon. These materials were used without further purification. The water was distilled and deionized.

Results

The percent recovery is shown at three solute loadings for toluene, in Figure 2, and for amyl acetate, in Figure 3. The recovery of toluene at three different flow rates is shown in Figure 4; that for amyl acetate is shown in Figure 5. The recovery of toluene at three different SDS concentrations is shown in Figure 6; that for amyl acetate at two different SDS concentrations in Figure 7. The percentage recovery of surfactant during the water flush step is shown at two flow rates in Figure 8.

Discussion

Effect of Solute Loading. In considering the results
in Figure 2 and 3 one naturally questions why the
percent recovery under some conditions is greater than
100 percent. When the solute loading was varied, the
most concentrated loading was used first, followed by
subsequent lower loadings on the spent carbon. Since
the purpose of these experiments was to investigate
the effect of solute loading, the regeneration was not
performed long enough to remove all of the solute.
For example, the base case run for the toluene was at
a solute loading of 0.1 mL/g carbon. If this run is
repeated a number of times, each regeneration
approaches 100 % recovery, even though a small level
of unrecovered solute remains on the bed. However,
this level of unrecovered solute is constant from run
to run. When the loading is substantially reduced,
this unrecovered solute at the end of a normal
regeneration (approximately 47 pore volumes) is less
than the unrecovered solute at the end of the higher
loading run. Hence, more toluene is recovered during
regeneration than was loaded on the carbon. When low
loading are used, only a small residual solute level
from a previous run can lead to substantially greater
than 100% apparent recoveries. At the point at which a
regeneration is terminated, though, the residual
solute is generally small, estimated to be 10% of that
loaded. The important conclusion from Figures 2 and 3
is that SECR is capable of effectively regenerating
carbon containing a range of loadings. The greater
than 100% recovery is an artifact of the experimental
procedure and the sequence of the experiments.

Effect of Multiple Cycles. The columns were operated
for a total of nine cycles. In order to test for
degradation of the carbon due to SECR after this many
cycles, some qualitative results were obtained.
Specifically, no solute was observed in the condenser
when the organic was being loaded on the carbon in any
of these experiments. If substantial reduction in
adsorption capacity were occurring, not all of the
organic could be adsorbed by the carbon. Therefore,
we conclude that this novel regeneration method did
not have serious deleterious effects on the
performance of the carbon in cleaning up volatile
organics from vapors over 9 cycles of operation.

Effect of Solute Type. Comparison of Figures 2 and 3,
4 and 5, and 6 and 7 shows that amyl acetate is more easily
recovered from the carbon than toluene (a given %
recovery is attained in fewer pore volumes). Recall
that the first step in the regeneration process is the
water flood, to displace air in the pore, to prevent
foaming. Therefore, the first several data points for

the first pore volume reflect the organic solute concentration in pure water in equilibrium with the organic solute on the carbon surface. This toluene concentration varied between 3.0 and 6.0 mM during this period, while the amyl acetate concentration varies between 15 and 20 mM. This greater tendency of the amyl acetate to distribute itself in the water phase caused it to be more easily removed from the carbon by the regenerant solution. It is also well known that solutes with aromatic rings solubilize at higher levels in micelles composed of cationic surfactants than in micelles composed of anionic surfactants (e.g., SDS), due to π electron attractions for the cationic head group and repulsion from the anionic head group (12,13). This could explain the preferential solubilization of the amyl acetate over the aromatic toluene in the anionic surfactant.

Effect of Regenerant Solution Flow Rate. From Figures 4 and 5, the greater the regenerant solution flow rate, the more pore volumes are required to regenerate the bed to a specified level. However, the effect is small for the toluene and much larger for the amyl acetate. If the regeneration step were equilibrium limited, flow rate would have no effect (when compared on a pore volume basis). Therefore, we conclude that mass transfer effects are significant for both solutes, but that the toluene is mainly limited by equilibrium solubilization considerations, while mass transfer effects are important for the amyl acetate. Of course, a higher flow rate could result in a shorter regeneration time, even if the recovery is poorer on a pore volume basis, hence requiring more total regenerant solution. If the regenerante were being recycled after removal of the volatile, this would not be a serious obstacle to using a higher flow rate.

Effect of Surfactant Concentration in Regenerant Solution. From Figure 6, the number of pore volumes required to achieve a specified percent recovery of toluene is approximately halved when the surfactant concentration in the regenerant solution is doubled from 0.1 M to 0.2 M. This is expected if the process is equilibrium limited and the solubilization equilibrium constant is independent of solute concentration. However, the surfactant concentration has little effect on the removal of amyl acetate, as seen in Figure 7. These results are consistent with those from the flow rate data: the toluene removal is mainly equilibrium limited while that of amyl acetate demonstrates significant mass transfer resistances.

Causes of Mass Transfer Limitation. Since amyl acetate has a higher molecular weight and is geometrically larger than toluene, it is expected to have a lower diffusivity. It could also have a higher activation energy for desorption from the carbon. Therefore, the results observed are reasonable, but more solutes need to be studied before drawing mechanistic conclusions.

Water Flush Step. As seen in Figure 8, flow rate had little effect on the number of pore volumes required to flush the surfactant from the carbon following the regeneration step. Therefore, under these conditions, the water flush step is equilibrium limited. In this case, if cycle time needs to be minimized, a large flow rate should be used; if not, a small flow rate will minimize downstream unit size to treat this water flush stream.

Conclusions

The overall conclusion of this study is that SECR can recover either toluene or amyl acetate from activated carbon over a range of loadings, over a range of regenerant solution surfactant concentrations, and with a reasonable number of pore volumes of regenerant solution. The basic feasibility of the process has been demonstrated.

Literature Cited

1.   Kovach, J.L. In "Handbook of Separation Techniques for Chemical Engineers"; P.A. Schweitzer, Ed.; McGraw-Hill, New York, 1979; Section 3.1.

2.   Hutchins, R.A. In "Handbook of Separation Techniques for Chemical Engineers"; P.A. Schweitzer, Ed.;   New   York,   1979; Section 1.13.

3.   Calgon Corporation, "Adsorption Handbook", Activated   Carbon   Division-Calgon   Corporation, Pittsburgh.

4.   Soffel, R.W., In "Kirk-Othmer Encyclopedia of Chemical Technology", 3rd Edition, Vol. 4, Wiley, New York, 1983; Vol 4, p. 563.

5.   Scamehorn, J.F. Ind. Eng. Chem. Process Des. Devel., 1979, 18, 210.

6.   Sutlkno, T.; Himmelstein, K.J. Ind. Eng. Chem. Fundam., 1983, 822, 420.

7.   Wankat, P.C.; Partin, L.C.; Ind. Eng. Chem. Process Des. Devel., 1980, 19, 446.

8.   Martin, R.J.; Ng, W.J. <u>Water Res.</u> 1987, 21, 961.

9.   Posey, R.J.; Kim, B.R. <u>J. WPCF</u> 1987, 59, 47.

10.   Chudyk,   W.A.;   Snoeyink,   V.L.   <u>Environ.   Sci.
Technol.</u>, 1984, 18, 1.

11.   Blackburn,   D.L.,   and   Scamehorn,   J.F.,
In "Surfactant-Based Separation Processes" Scamehorn,
J.F.;   Harwell,   J.H.,   Eds.;   Marcel   Dekker,   New
York, In Press.

12.   Lianos,   R.;   Viriat,   M.L.;   Zana,   R.   <u>J.
Phys. Chem.</u>, 1984, 88, 1098.

13.   Almgren,   M.;   Grieser,   F.;   Thomas,   J.K.   <u>J.
Am. Chem. Soc.</u>, 1979, 101, 279.

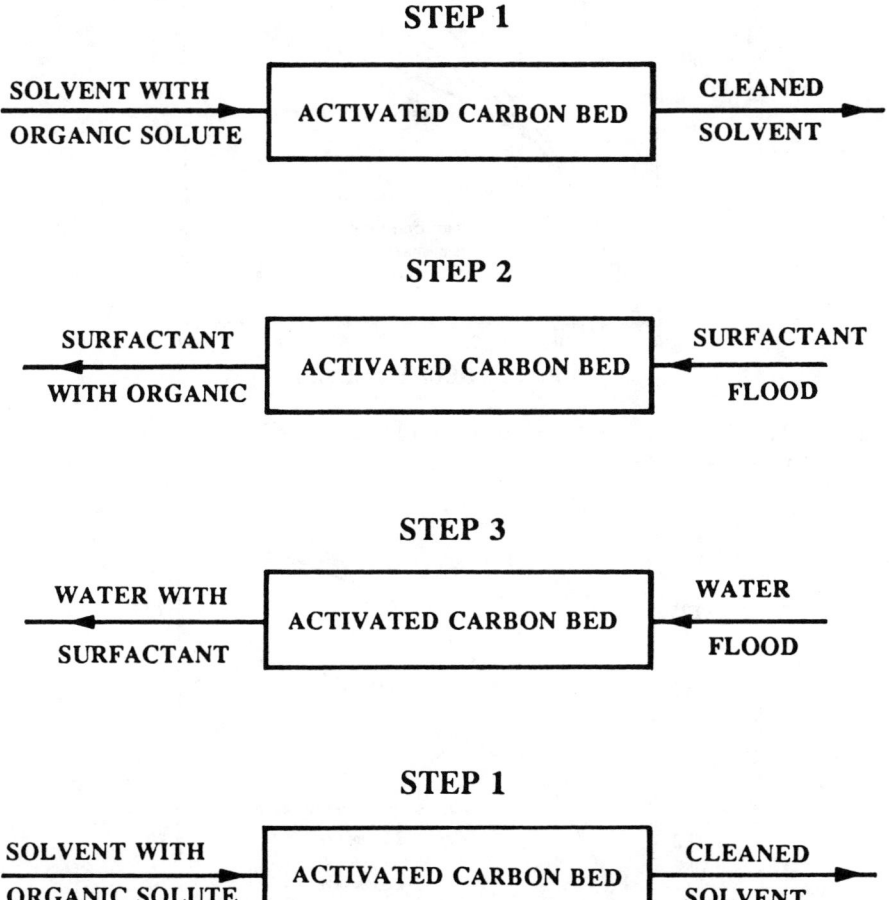

Figure 1.   Surfactant Enhanced Carbon Regeneration Process.

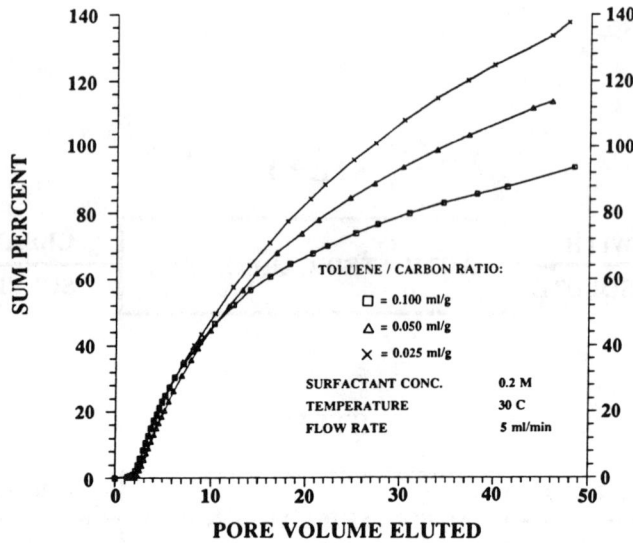

Figure 2. Effect of toluene loading on the recovery of toluene.

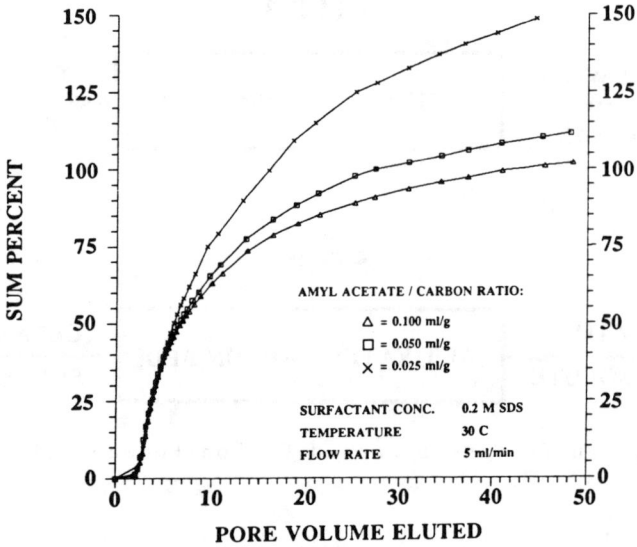

Figure 3. Effect of amyl acetate loading on the recovery of amyl acetate.

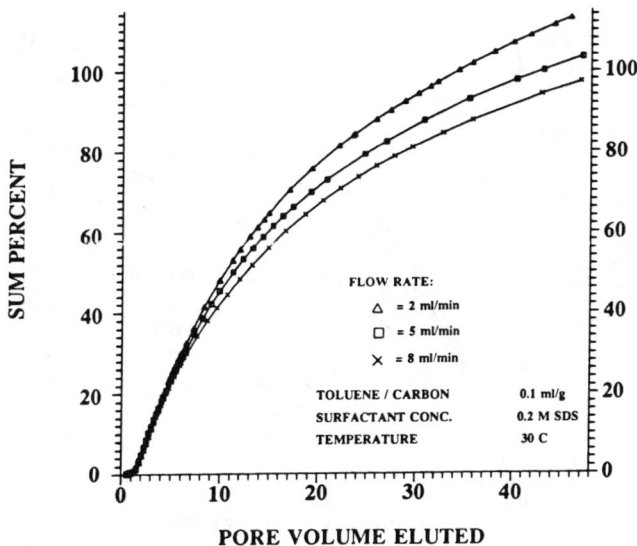

Figure 4. Effect of surfactant solution flow rate on the recovery of toluene.

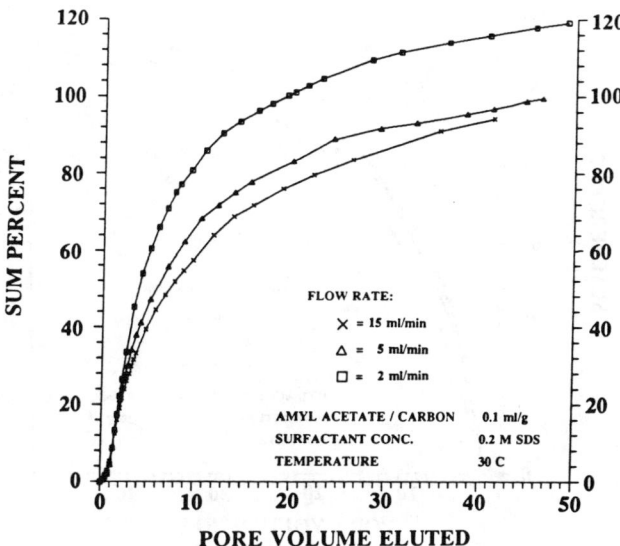

Figure 5. Effect of surfactant solution flow rate on the recovery of amyl acetate.

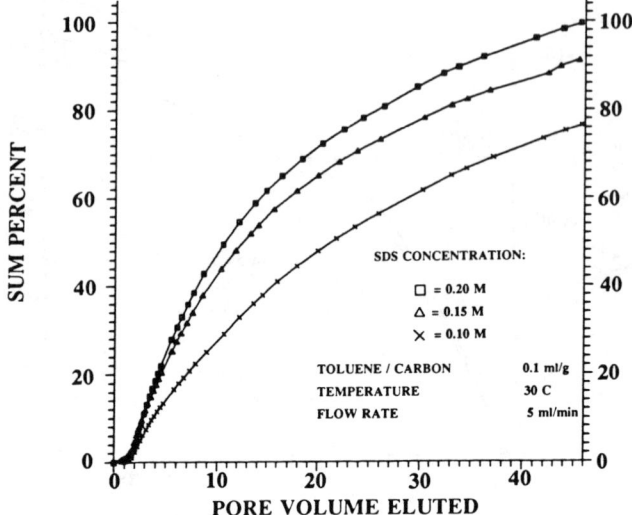

Figure 6.   Effect of surfactant concentration on the recovery of toluene.

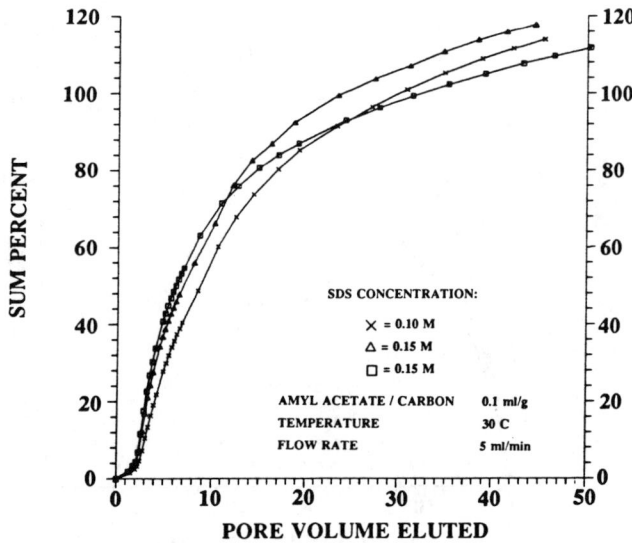

Figure 7.   Effect of surfactant concentration on the recovery of amyl acetate.

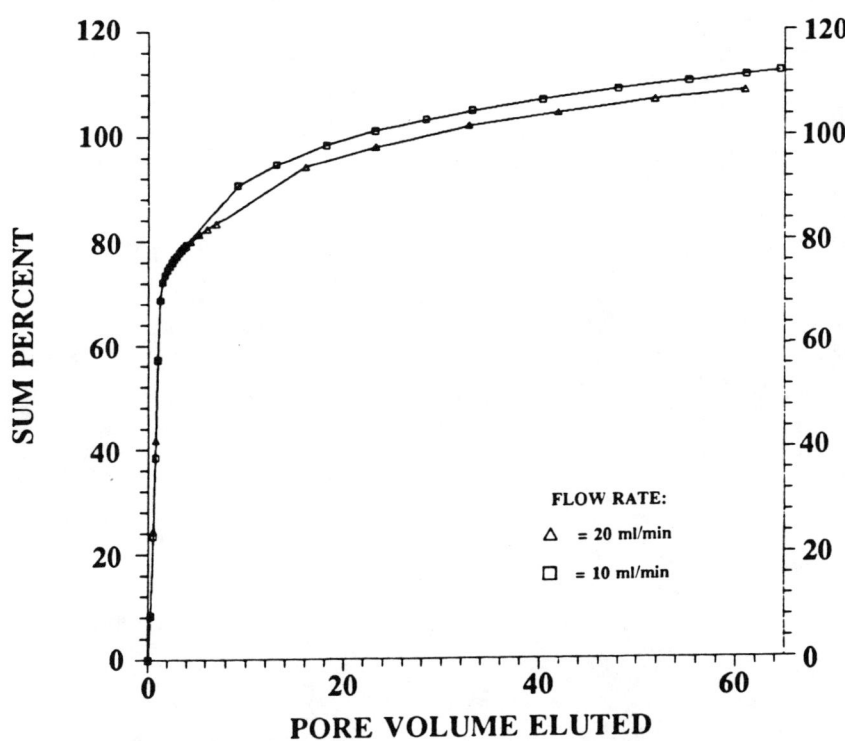

Figure 8. Effect of water flow rate on the recovery of residual surfactant.

Chapter 9

Developments in Batch and Continuous Liquid Chromatographic Separations

P.E. Barker

G. Ganetsos

The scaling up of batch and continuous chromatographic systems from column diameters of 5.1 cm to 10.8 cm and total length of over 700cm is reported. The fractionation of the macromolecule dextran has been carried out successfully by employing the size exclusion principle to remove the higher and lower molecular weight fractions, and obtaining a dextran product of narrower molecular weight range. Synthetic glucose-fructose mixtures, inverted sucrose, industrial carbohydrate effluents and industrial barley syrups were separated successfully by operating in the ion-exchange mode. High fructose syrups (HFCS) of the highest specification were produced. The switch time was found to be the controlling parameter in the operation of the semi-continuous chromatographic refiners (SCCR).

INTRODUCTION

Chromatography is a low energy intensive unit operation of high separation potential. It does not involve any phase change, and compounds of very close boiling points, isomers and solute mixtures with low separation factors have been separated efficiently. The separation is achieved due to the different migration rates of the solutes through a system of two phases, the mobile phase and the stationary phase.

The great separation potential and the versatility of chromatography have been widely accepted and chromatography has been established as a powerful analytical tool. Although over the last twenty years or so industry has seen the employment of some large scale processes, especially in the carbohydrate, petrochemical and essential oil fields, it is our belief that chromatography's large scale separation potential has not yet been fully exploited.

The scaling up of such processes in the 60's and 70's was based towards the batch mode with the notable exception of one or two continuous systems. There are in principle two alternative approaches in achieving continuous chromatographic operation, namely "cross-current" and "counter current".

In cross-current chromatography the mobile phase moves almost at right angles to the stationary phase. The mobile phase is introduced continuously at the top of the annulus and leaves at the bottom. The components travel in helical paths around the annulus at different angles according to their relative affinities for the stationary phase and are eluted at different points at the bottom of the column [1]. Theoretically, the continuous separation of a multicomponent mixture can be achieved using such a configuration.

A number of continuous counter-current chromatographic processes have been developed and can be classified according to the principle they employ to achieve the counter-current movement. In the moving bed systems the stationary phase flows under gravity counter-current to a stream of mobile phase flowing upwards. When the feed mixture is introduced into the column, usually in the middle, the retarded component is carried with the packing and is stripped at the column outlet and the least absorbed component is carried upwards with the mobile phase [2].

The moving column continuous systems consist of a circular array of interconnected columns, which rotate as a whole past fixed inlet and outlet ports. A number of alternative schemes have been employed and some systems have been developed with considerable success in the analytical field [3].

In the moving port systems a number of static columns or compartments are interlinked and the countercurrent movement is effectively achieved by sequentially moving the inlet and outlet ports associated with each column, in the direction of the mobile phase.

There are two main approaches: the Sorbex process developed by the Universal Oil Products (USA) [4], and consisting of a single column divided into a number of compartments where the inlet and outlets in each compartment are controlled using a master multiport valve, and the semi-continuous-chromatographic-refiners (SCCR) which were developed in this department and will be described in detail later.

This paper reports recent developments in employing batch and continuous (SCCR) chromatographic preparative scale systems in the fractionation of macromolecules, separation of carbohydrate synthetic and industrial feedstocks and in the production of high fructose syrups (HFS). In the fractionation experiments the size exclusion principle was employed, while the separation of carbohydrate mixtures was achieved using the ion-exchange principle.

SEMI-CONTINUOUS CHROMATOGRAPHIC PRINCIPLE OF OPERATION

The SCCR system consists of a number of interconnected columns, with six valves attached to each column. At the column inlet, the eluent, feed and purge valves are attached while at the column outlet there are two product outlet valves and a transfer valve on the flow line connecting two adjacent columns together. Each of these valves are connected to the corresponding eluent, feed, purge and two product distribution pipe networks. For illustration purposes a twelve column system is considered and in Figure 1 the whole system is represented as a closed loop. The feedstock is introduced continuously at port F, and the mobile phase flows continuously through to port M. The less strongly adsorbed component B moves preferentially with the mobile phase towards the product outlet P1.

A section of the loop is isolated at any time by the two transfer valves V1 and V2, and an independent purge stream enters at port PU, strips the adsorbed component A and flows out of port P2.

After a predetermined time interval, referred to as the "switch-time" all the valve (or port) functions advance by one position in the direction of the mobile phase flow, thus achieving the countercurrent movement. This is illustrated in

Figure 2. Figure 2a represents the first switch period where column 1 is isolated and purged to give the product rich in A. Feed and eluent enter columns 7 and 2 respectively and the product rich in B is eluted from column 12. In the next switch period all ports are advanced by one position as shown in Figure 2b and so on. After twelve such advancements a "cycle" is completed, and after approximately six cycles the concentration profiles in the system became reproducible and a "pseudo-equilibrium" state is reached. The operating principle has been reported in detail in reference (3).

To achieve a separation and hence two enriched products, the rate of port advancement must be greater than the migration velocity of component A through the bed and lower than the migration velocity of B. The theoretical limits of the mobile and stationary phase rates of movement are represented by:

$$K_{dB} < \frac{Lm}{P} < K_{dA} \tag{1}$$

A detailed theoretical explanation can be found in references 3 and 5.

PREPARATIVE SCALE GEL PERMEATION CHROMATOGRAPHY

The fractionation of polymers according to their relative molecular sizes is well established on the analytical scale by employing gel permeation chromatography (GPC).

As a brief description of the separation principle, when a polymer solution flows through a bed packed with a porous solid, the smaller molecules penetrate the pores completely and are eluted last at a volume corresponding to the total liquid volume of the column. The larger molecules are excluded totally from the pores and are eluted first at a volume equal to the interstitial column volume, while the medium size molecules penetrate the pores to a certain extent and are eluted at intermediate volumes. The polymer used in the preparative scale fractionation studies was dextran, a polyglucose used mainly in medicine as a blood plasma volume expander.

The various preparative GPC systems used were packed with Spherosil XOB075 of 200 to 400 μm porous silica beads, employing the slurry packing technique. Spherosil was chosen for its compatibility with dextran/water systems, its relative incompressibility, and wide availability of different size ranges. The dextran molecular weight of interest was between 10000 to 150000 daltons requiring the most suitable Spherosil XOB075 as chromatographic media. In selecting the particle size range, factors such as HETP, pressure drop and cost were considered. Ellison (6) has found that the HETP values were decreased on average by a factor of three when the Spherosil XOB075 size range was changed from 200-400 μm to 95-105 μm.

Glucose was used as the test solution at eluent flow rates ranging from 0.5 to 5.5 cm^3 min^{-1}. Due to the viscous nature of the dextran however it was decided to use the 200-400 μm packing in the preparative scale since the pressure drop per unit length increases in proportion to the square of the particle diameter. In a latter study Spherosil XOB30 was also used. Initially (SCCR3 and batch results) the molecular weight distribution of the dextran samples from the preparative systems were determined by GPC analysis using a 0.4cm dia x 120cm long glass column packed with Porasil C of very close column size range

(95-105μm). Later (SCCR5 results) the analytical GPC system consisted of one PW3000 and two PW5000 TSK columns.

The various experimental runs carried out will be reported in a coded form, ie. as 23.1-40-101.1-7.5-20 for the continuous and 16.9-102-100-20 for the batch. In the continuous mode the notation of the figures indicate feed concentration (% w/v), feed and eluent flow rates (cm^3/min), switch time (min) and operating temperature (°C), while in the batch mode the notation used is feed concentration (% w/v), feed charge (cm^3), eluent flow rate (cm^3/min) and operating temperature (°C) respectively.

<u>Continuous Operation</u>

Figure 3 represents typical dextran on column concentration and molecular weight distribution profiles on the SCCR3 equipment when pseudoequilibrium has been established (run 21-6.9-43.9-17.6-20). As it is apparent, on-column concentrations up to 16% w/v (for a 21% w/v feed concentration and the corresponding operating conditions) exist and the sequential fractionator still operates successfully. Analytical batch GPC columns usually operate at concentrations lower than 1% w/v to maintain resolution so that the 16% w/v on column concentrations represent severe overloading by comparison. This indicates a particular advantage of the continuous countercurrent systems, since high throughputs and increased product concentrations are required in production scale operations. From the above figure it is also apparent that the major change of molecular weight distribution, expressed in terms of K_d values, occurs in less than half of the total length available for separation. Originally (7) it was believed that by moving the location of the feed inlet in the direction of the "plateau" part of the profile, the effective system length available would be increased. Although this is correct in principle it requires reprogramming of the pneumatic controller covering the functioning of the various inlet and outlet valves. As will be seen later on this can be achieved instead by simply adjusting the switch-time.

The fractionation achieved can be seen visually by comparing the feed and product distribution profiles in Figure 4, where the areas of the product profiles are expressed in proportion to the eluted dextran mass rates (run 21-7.9-43.9-17.6-20). A selection of the experimental results obtained on the SCCR3 system is shown in Table II.

The results for run 21-7.9-43.9-17.6-20 show clearly the fractionation achieved and substantiate the graphical results as shown in Figure 4. The feed used in this run was Dextran 40 (Fisons Pharmaceuticals plc, batch BT85). All the other runs reported in Table II were also carried out with Dextran 40 (Fisons Pharmaceuticals plc, batch BT192) which had a substantially different molecular weight distribution.

A comparison between runs 1.02-10.1-51.7-16.9-20 and 22-10-52.6-17-20 indicate the effect of increasing the feed concentration whilst maintaining the other operating conditions the same. Although there is a reduction in fractionation performance at higher feed concentrations the fractionation achieved is still satisfactory. The increased feed concentration results in a reduction in solute mass transfer rates and an increase in solute zone broadening. This reduction in fractionation can be compensated for by the resultant increases in throughput and product concentrations and in an industrial application an overall economic evaluation will optimise the above parameters.

In runs 22-7.22-37.8-30-20, and 22-10-52.6-17-20 the feed flow rate was increased while the eluent to feed flow ratio and feed concentration were kept the same. The results indicate that in fact it is advantageous to operate at higher feed and eluent flow rates since better fractionation is obtained at higher throughputs.

Although this result appears to be very encouraging one should accept it with some scepticism. This increased fractionation efficiency might be due to the different switch time used. While everything else was kept the same the switch time was not changed in proportion to the flow rates. This means that the corresponding "cut positions" were different, the two runs were carried out at different linear velocities and the linear velocity used in run 22-10-52.6-17-20 was the more appropriate. This effect of switch time changes has been identified and investigated more fully in the continuous separation of barley syrups using the SCCR7 system and will be explained later on.

All the results reported in Tables III and IV were carried out on the SCCR5 system and the operating conditions were geared in such a way to study separately the removal of either the HMW or the LMW dextran fraction present in the feed. This is illustrated in Figure 5 (8, 9) which shows the respective on-column concentration profiles of runs 23.1-40-101.1-7.5-20 and 23.6-17.5-208.8-5-20. In the profile of run 23.1-40-101.1-7.5-20 most of the dextran feed moves with the stationary phase leaving a small amount of high molecular weight dextran to be eluted with the mobile phase, whilst in the profile of run 23.6-17.5-208.8-5-20 most of the dextran feed is eluted with the mobile phase and only a small amount of LMW dextran is removed by being retarded on the stationary phase and eventually purged. The corresponding $\overline{M}_w$ and $\overline{M}_n$ values in Table IV reflect the above.

The effect of temperature becomes apparent in the results of runs 21.5-10.7-52.1-20-20 and 23.2-10.4-52.1-20-45 where a 25°C rise, keeping everything else almost the same, resulted in a 30% reduction of the amount of dextran removed in the LMW product. The resulting differences in the $\overline{M}_w$ and $\overline{M}_n$ values also confirm this.

Experience so far has shown that the determination of the "cut-positions" has a dominant effect on the fractionation achieved,while temperature increases change the quality and quantity of the dextran held-up in the fractionating section and have a slight effect on the product split. Increases in the operating temperature have more prominent effects when the experiments are geared towards the removal of the LMW fractions in the dextran feed.

The setting of the operating cut-positions and thus the fractionation was controlled by adjusting the respective eluent and feed flow rates. This had a direct impact on the product concentration, and the maximum concentration of the main product was just over 3.5% w/v. Although the actual dextran feed used in the SCCR5 experiments was not from the same batch, similar feed types were used and had approximate $\overline{M}_w$ and $\overline{M}_n$ values of 64300 and 27000 respectively. In the experiments geared to removing the LMW fraction of the feed stream, ie. runs 23.6-17.5-208.9-5-20, and 21.5-10.7-52.1-20-20, the aim was to increase the $\overline{M}_w$ by more than 10% and the $\overline{M}_n$ by more than 1000 (Tables III and IV). As it is apparent from the results this was clearly achieved. The objective, in carrying out run 23.1-40-101.1-7.5-20, was to reduce the $\overline{M}_w$ and $\overline{M}_n$ values of the LMW product substantially by removing no more than 15% of the dextran in the feed as the HMW product. This was achieved successfully and the $\overline{M}_w$ and $\overline{M}_n$ values in the LMW product were 49300 and 23700 respectively.

To overcome the pressure limitations the glass columns were replaced by similar sized stainless steel columns (SCCR5 mk 2) and the same packing material being used. The experiments were focused on removing the HMW from the feed and producing a dextran LMW product with $\overline{M}_w$ of less than 33000. Similar dextran feeds were used with $\overline{M}_w$ and $\overline{M}_n$ values of 82000 and 9000 respectively and the operating temperature being kept at 60°C. The experimental operating conditions and results are shown in Table V. The results show that a reduction in the feed flow rate (runs 13.4-60-95-7.5-60 and 13.1-40-95-7.5-60) alters the actual "cut position" resulting in a reduced removal of HMW dextran ($\underline{10}$, $\underline{11}$). The increased % dextran recoveries and $\overline{M}_w$ value give support to this finding.

Since stainless steel columns were now used, pressure was not a limiting factor any more, therefore the columns were packed with Spherosil XOB030 to investigate any changes in the fractionation efficiency. Run 13.1-60-96-7.5-60 was then carried out and the operating conditions were kept the same as in run 13.4-60-95-7.5-60 which was carried out using the XOB075 packing. Although the $\overline{M}_w$ values obtained were still within the specification set above, the amount of HMW dextran removed in the HMW product was reduced, resulting in slightly higher % recoveries and $\overline{M}_w$ values in the desirable LMW product stream. This indicated that the Spherosil XOB075 packing was more suitable for the removal of the HMW dextran in the dextran feedstock. Finally it was found that the % of dextran removed as HMW product increases as the eluent flow rate increases, as illustrated by the results of runs 7.1-25-116-7.5-60 and 7.1-25-125-7.5-60.

<u>Batch Operation</u>

The batch experiments were carried out using the batch chromatographic refiner (BCR1) which consisted of the ten borosilicate columns employed in the SCCR5 system. To minimise the "dead" volume the valves were bypassed and the columns were interconnected directly. All experiments were carried out at ambient temperature and using the same dextran feed, having $\overline{M}_w$ and $\overline{M}_n$ values of 47500 and 20800 respectively. The product specification had a $\overline{M}_w$ value of less than 40000, a M_{10} value greater than 13000 and a M_{90} value of less than 130000 respectively. The repetitive feed injection technique was employed to maximise throughput. A selection of the results obtained is shown in Table VI, from which it becomes apparent that the fractionation of dextran to the above specification can be achieved.

Both the very high and the very low molecular weight material was removed from the injected feed in a single pass through the batch system and produced fractions having a narrower molecular weight distribution. The maximum injected volume that gave material within specification was found to be equivalent to 3.3% of the total liquid volume of the column ($\underline{8}$, $\underline{9}$). An increase in the feed charge (runs 16.9-102-100-20 and 17.2-226-100-20) gave products having broader molecular weight range and higher molecular weight averages. Similar effect is brought about by increasing the feed concentration and keeping the feed charge constant (runs 17.2-226-100-20 and 26.3-239-100-20, also runs 16-149-100-20 and 26.8-145-100-20).

<u>PREPARATIVE ION-EXCHANGE CHROMATOGRAPHY</u>

In the scaling up work of the batch and continuous SCCR systems the separation of carbohydrate mixtures was employed due to the low hazards in handling sugar solutions instead of say hydrocarbons and also due to the increased uses of High Fructose Corn Syrups (HFCS) as a sweetener. This demand became apparent early in the 1970's after the reshaping of the American carbohydrate market, when a number of large batch chromatographic systems were introduced. A recent report in the Times newspaper ($\underline{12}$) predicted that the commercial output of maize-based crystalline fructose will double in the USA after the opening of a new production plant. HFCS are potentially the biggest threat to sugar's share in the market, and now have about 40% of the total USA market all of which has happened since the beginning of the 1970s. The corn syrup is produced from locally grown maize.

In this work a number of alternative feedstocks were used such as synthetic glucose-fructose mixtures. The industrial effluent from a dextran production plant, inverted sucrose and industrial barley syrups. Both anion and cation exchange resins were employed.

Based on the information available all the industrial HFCS producing chromatographic systems employ cation exchangers, usually polystyrene resins in the calcium form (Ca^{2+}) and of various degrees of cross-linking. The fructose in the carbohydrate feedstock forms a loose complex with the calcium ions and is retarded, whilst the glucose and other oligosaccharides are eluted with the mobile phase (13). Although some analytical chromatographic columns are known to operate with anion-exchange it is understood that no industrial systems use anion exchangers. In anion-exchange chromatography the resin is usually in the bisulphite (HSO_3^-) form where the glucose is retarded by complexing with the HSO_3^- whilst the fructose and other oligosaccharides are eluted first.

Continuous Chromatography

The SCCR7 mk1 system was used in the anion-exchange mode to investigate factors affecting the separation of synthetic equimolar glucose-fructose and hydrolysed sucrose feedstocks. The operating conditions and results are shown in Tables VII and VIII, and the first experiments showed that it was possible to achieve 99% pure fructose rich product from an equimolar feedstock.

An increase in the feed flow rate (runs 51.6-7-28-30-60 and 49.9-10-28-30-60) disturbs the criteria for separation (equation 1) by altering the Lm/P ratio, and has been found to shift the on-column concentration profile towards the FRP (14, 15). This indicates that an increased amount of glucose now travels with the mobile phase reducing the FRP purity (from 99 to 84%). An increase of the eluent to feed ratio from 4 to 4.3 (runs 51.6-7-28-30-60 and 50-7-30-30-60) also alters the Lm/P ratio, but in this case it results in higher recoveries of fructose in the FRP at a slight expense of the FRP purity.

To investigate the effect of changing the source of feedstock runs 50-7-30-30-45 and 49.1-7-30-30-45 were carried out at identical conditions. In the latter run however an inverted sucrose solution was used as a feedstock. The sucrose was hydrolysed by passing it through a 5.08cm id x 75cm long glass column packed with Amberlite IR118 resin in the H^+ form the inverted product being fed directly to the SCCR7 mk1 system. The result shows that there is no difference in the separation performance of the SCCR7 mk1 system and hence sucrose feedstocks such as molasses can be inverted and separated. Although the use of molasses appears to be theoretically possible, the effects of the other organic and inorganic impurities on the separation and the corrosion problems due to the ionics present need to be investigated and overcome first.

Due to the nature of the operation of the SCCR system the desorption of the retarded component needs increased flow rates resulting in dilute products. If the desired product is fructose then the anion exchange principle is advantageous because higher fructose product concentrations are possible since glucose is the retarded component. When however a multicomponent carbohydrate feedstock is used (more than 2 components) then the oligosaccharides present migrate faster since they are excluded due to their larger molecular size; fructose travels slower than the oligosaccharides whilst the glucose is retarded due to ion complexing. Therefore, it is not possible to separate fructose in a "single" pass since now it becomes the "middle" component and can be contaminated by products on either side. In addition to that, anion-exchange resins are known to suffer from auto-oxidation brought about by the dissolved oxygen in the eluent, this being enhanced by the presence of Fe^{2+} and Cu^{2+} ions (16) and also by mechanical fouling (17). Therefore cation exchange resins were employed in subsequent experiments.

To investigate the effect of scaling up, the SCCR6 system was used having 10.8cm diameter columns, and an equimolar glucose-fructose feed was used (18, 19). At low feed concentrations pure products were obtained (20-35-105-30-20), but as the feed concentration was increased the product purities were affected, Table X. As the feed concentration was increased more glucose moved with the stationary phase resulting in a shift of the glucose on-column concentration profile towards the FRP, as illustrated in Figure 6. The SCCR6 system was also used to recover the fructose from a fructose rich dextran effluent containing 70% w/v sugar solids which contained 69.6% fructose, 21.5% dextran, 8.9% glucose and other reducing sugars. Operating at a throughput of 2.94 kg/h or 48.6 kg sugar solids/m^3 resin/h, pure FRP was obtained at a 16.3% w/v concentration. The FRP was split into two fractions the dilute part being recycled as eluent.

The majority of the previous experiments were carried out using feedstocks with fructose to glucose ratios of 50:50 or higher. In the following experiments an industrial isomerose feedstock consisting of about 42% fructose, 52% glucose and 6% maltose and other oligosaccharides was used. The objective was to produce an FRP of over 90% purity and a GRP containing less than 7% fructose at a feed throughput of over 30 kg sugar solid/m^3 resin/h with product concentrations of over 20% w/v. The 5.4cm id SCCR7 mk2 (20, 21) system was used in this work because of the reduced feed and eluent volumes needed and also because it allowed continuous unattended operation.

In the results obtained so far the effects of the various operating parameters such as flow rates, eluent to feed ratios, feed concentration etc on the separation performance were identified and although the FRP purity was kept high it was achieved at the expense of the overall fructose recovery, consequently increasing the GRP fructose content. In the above experiments and also in the dextan fractionation work the "cut-positions" were set by trying different eluent and feed flow rates and thus altering the Lm/P ratio. The results of runs: 36-13-40-23-60 and 37-13-40-25-60 (Tables X1 and XII) however show that this can be achieved by simply altering the switch time while maintaining everything else the same. This is illustrated in Figures 7 and 8 which shows the shift of the on-column concentration profiles brought about by the switch time changes. The switch time is the controlling parameter in the operation of the semi-continuous chromatographic systems, whether they are used either as fractionators or separators.

The problems associated with feed concentration increases, ie. the shift of the glucose concentration profile towards the FRP, can be overcome by selecting the right switch time and hence artificially shifting the profile in the opposite direction. This is illustrated by the results of runs: 36-13-40-23-60, 46-13-39-24.17-60 and 54-13-39-24.5-60. The increased sugar concentrations increased the separation difficulty and although the product purities were reduced slightly they were still within specification (21).

Previous experience had shown that the minimum eluent to feed flow rate ratio was 3 to 1. To evaluate the possibility of reducing it further, run 47.5-13-32-28.5-60 was carried out at an eluent to feed ratio of 2.46 to 1, and with the purge flow rate reduced to 60 cm^3 min. Comparing the results with that of run 46-13-39-24.17-60 where a 3 to 1 ratio was used it can be seen that this reduction resulted as expected in higher product concentrations but at the expense of their purities.

It was therefore concluded that a value of 2.75 to 1 is the absolute minimum recommended eluent to feed ratio, and the minimum purge flow rate should not fall below 70 cm^3/min.

It has been found that most of the fructose in the FRP is eluted in the first half of the switch period and most of the glucose in the GRP is eluted in the second half of the switch period. Therefore by collecting only the concentrated portions the product concentrations can be increased considerably. To increase the product concentrations further the feed concentration was increased to 66%w/v, the eluent to feed flow rate ratio reduced to 2.74 to 1 and the purge rate maintained at 70 cm^3/min (run: 66-14.6-40-25-60). The purities of both products were well within specification, and the concentrations of the collected portions were 11.29 and 22.56% w/v for the FRP and GRP respectively. Over 94% of the glucose in the feed was recovered in the GRP and almost 96% of the fructose in the feed was recovered in the FRP, at a throughput of 32.1 kg sugar solids/m^3 resin/h.

To minimise the eluent requirements however and recover all the sugar solids entering the system it was decided to recycle the dilute product splits as eluent and purge water (run: 66.3-14.6-40-26.5-60). This approach increased the FRP and GRP concentrations to 13 and 25.4% w/v respectively whilst both product purities were still within specification; and the total carbohydrate throughput was 32.3 kg sugar solids/m^3 resin/h.

Batch Operation

The columns in the SCCR5 system were interconnected directly by bypassing the valves to convert it in to a long batch system (BCR2). Synthetic equimolar glucose-fructose mixtures were used with concentrations varying from 20 to 60% w/v. The eluent flow rate was kept the same to the one used in the continuous SCCR6 runs, ie. 105 cm^3/min. In all experiments the overlapping part of the elution profiles was recycled in order to obtain pure glucose and fructose rich products. An increase in the feed concentration runs 20-15000-105-20, 40-15000-105-20 and 60-15000-105-20 (Table XIII), resulted in increased throughputs and product concentrations, up to 15.15 and 21.85% w/v for the FRP and GRP respectively.

The effects of increasing feed charges are shown by the results of runs 20-1200-105-20, 20-6000-105-20 and 20-15000-105-20. Due to feed band broadening effects (22) it is recommended to use higher feed concentrations instead of increased feed volumes. The maximum throughput achieved was 1.5 kg/h or 24.8 kg sugar solids/m^3 resin/h, which was almost 23% lower than that achieved on the continuous SCCR7 mk2 when it was used to separate barley syrups. The batch operation at high feed concentrations produced pure products but required substantial recycling.

When 12 litre charges of the 69.4% w/v dextran effluent feed were injected the DRP purity was 66%, the FRP 99.9%, and the respective product concentrations were 3.84 and 8.64% w/v. To achieve the above FRP purity however required 48% in weight terms recycle of the total sugar solids injected. This resulted in a throughput of 0.56 kg/h or 9.26 kg sugar solids/m^3 resin/h which was far below the 48.6 kg sugar solids/m^3 resin/h achieved when it was operated continuously using the SCCR type system (Table X).

CONCLUSIONS

Both batch and semicontinuous chromatography have been found to operate efficiently in the fractionation of macromolecules such as dextran, and in the separation of carbohydrate mixtures.

The SCCR equipment was used successfully to separate industrial barley syrups and to produce high fructose syrups meeting the strictest specifications. Operating at throughputs of 32.3 kg sugar solids/m^3 resin/h the FRP was over 90%, its concentration was almost 13% w/v and the glucose rich product contained less than 6.7% fructose and had a concentration of over 25% w/v. The overall results indicated that at high feed concentrations of binary mixtures, the batch operation is slightly better in terms of product quality. The semi-continuous operation however, offers better throughputs, is more flexible, requires no recycling, allows continuous unattended operation and ensures constant product quality.

Although these advantages favour the application of continuous chromatography in production scale fractionation and separation processes, there are cases where the batch operation is superior and therefore the selection of the best mode of operation needs careful consideration for any mixture separation.

ACKNOWLEDGMENTS

The authors would like to thank Drs Ellison, England, Abusabah and Thawait for the provision of some of the experimental results.

LEGEND OF SYMBOLS

AEC	Anion Exchange Chromatography
BCR	Batch Chromatographic Refiner
CEC	Cation Exchange Chromatography
DRP	Dextran Rich Product
FRP	Fructose Rich Product
GPC	Gel Permeation Chromatography
HETP	Height Equivalent to a Theoretical Plate
HFCS	High Fructose Corn Syrups
HFS	High Fructose Syrups

K_{di} Distribution coefficient of component $i = \dfrac{\text{mass of i in stationary phase}}{\text{mass of i in mobile phase}}$

Le Pre-feed mobile phase flow rate

Le' Post-feed mobile phase flow rate

Lm Average mobile phase flow rate $= \dfrac{\text{Le+Le'}}{2}$

$\bar{M}_n$ Number average molecular weight

$\bar{M}_w$ Weight average molecular weight

M_{10}&M_{90} Molecular weights corresponding to the 10% and 90% points on the cummulative molecular weight distribution of the products

P Stationary phase effective flow rate $= \dfrac{V}{S}$

S Switch time

SCCR Semi-continuous Chromatographic Refiner

V Column volume

LITERATURE CITED

1 Canon RM, Begovitch JM, and Sisson WG, Separation Science and Technology, 1980, 15 (3), 655.

2 Berg C, Chem Eng Proc, 1951, 47, 585, 1951.

3 Barker PE,"Developments in Chromatography",Ch.2,Appl.Science, London,1978

4 Broughton DB, Chem Eng Progr, 64, 1968, (8), 60.

5 Barker PE & Ganetsos, Separation Science and Technology, Giddings JC, Gruska E, Eds. Dekker, New York, 1987, Vol. 22, No. 8-10, pp 2011-203.

6 Ellison FJ, PhD thesis, Aston University, Birmingham, 1976.

7 Barker PE, Ellison FJ and Hatt BW, Ind Eng Chem Process Des Dev, 1978, Vol 17, No 3, 902, 1978.

8 England K, PhD thesis, Aston University, Birmingham, 1979

9   Barker PE, England K, Vlachogiannis G, <u>Chem. Eng. Res</u>. Des., July 1983, Vol 61, 241.

10 Vlachogiannis G, PhD Thesis, Aston University, Birmingham, 1982

11 Alsop RM, Barker PE, Vlachogiannis G,<u>The Chemical Engineer</u>, 24, January,1984.

12 Times newspaper, London., "Long term threat to sugar price", 1 June 1987

13 Goulding RW, <u>J of Chromatogr</u>, 1975, <u>103</u>, 229.

14 Abusabah EKE, PhD thesis, Aston University, Birmingham, 1983

15 Barker PE, Abusabah EKE, <u>Chromatographia</u>, Jan. 1985, Vol. 20, No.1, 9.

16 Kunin R, <u>Ind Eng Chim</u>, 1959, <u>49</u>, 1365.

17 Cores AF, "Ion exchangers: properties and applications", Ann Arbor Science, Michigan, 1972

18 Thawait S, PhD thesis, Aston University, Birmingham, 1983

19 Barker PE, Thawait S, <u>Chem Eng Res Des</u>, July 1986, Vol 64, 302.

20 Ganetsos G, PhD thesis, Aston University, Birmingham, 1986

21 Barker PE & Ganetsos G, <u>J Chem Tech Biotechnol</u>, 1985, <u>35B</u>, 217

22 Barker PE,Ganetsos G & Thawait S,<u>Chem.Eng.Science</u>,1986,<u>41</u>,(10), 2595.

Table I - Batch and Continuous Chromatographic Systems

Equipment	Mode of operation	Separation principle	Stationary phase and size range	Mobile phase	Total system length (cm)	No of columns and material of construction	Column diameter (cm)
SCCR3	Continuous	GPC	Spherosil XOB 075 (200 to 400 μm)	Distilled water	700	Ten (glass)	5.1
SCCR5	Continuous	GPC	Spherosil XOB 075 (200 to 400 μm)	Distilled water	700	Ten (glass)	5.1
SCCR5 (mk2)	Continuous	GPC	Spherosil XOB 075 (200 to 400 μm)	Distilled water	700	Ten (st/st)	5.1
BCR1	Batch	GPC	Spherosil XOB075 (200 to 400 μm)	Distilled water	780	Ten (glass)	5.1
SCCR7 (mk 1)	Continuous	AIE	Duolite All3 (HSO$^-_3$ form)	Deionised water	780	Twelve (st/st)	5.4
SCCR6	Continuous	CIE	Zerolit SCR14 (150-300 μm)	Deionised water	660	Ten (st/st)	10.8
SCCR7 (mk 2)	Continuous	CIE	Korela V07C (150-300 μm)	Deionised water	780	Twelve	5.4
BCR2	Batch	CIE	Zerolit SRC14 (150-300 μm)	Deionised water	660	Ten (st/st)	10.8

where:
SCCR :	Semi-Continuous Chromatographic Refiner
BCR :	Batch-Chromatographic Refiner
GPC :	Gel Permeation Chromatography
CEC :	Cation Exchange Chromatography
AEC :	Anion Exchange Chromatography

Table II - Dextran fractionation using the SCCR3 system

Run	Flow rates (cm^3/min)			Eluent to feed flow ratio	Feed conc % w/v	Switch time (min)	Total No of cycles	Total time taken (h)	Mean molecular weights of the two products	
	Eluent	Feed	Purge						HMW	LMW
21-7.9-43.9-17.6-20	43.9	7.9	195	5.56	21.00	17.6	15	43.9	95000	27000
1.02-10.1-51.7-16.9-20	51.7	10.1	190	5.12	1.02	16.9	12	34.0	35000	16000
22-10-52.6-17-20	52.6	10.0	190	5.26	22.00	17.0	10	28.4	44000	26000
22-7.22-37.8-30-20	37.8	7.22	160	5.24	22.0	30.0	8	40.0	29000	28500

Key: Run code indicates feed concentration (% w/v), feed flow rate (cm^3/min), eluent flow rate (cm^3/min), switch time, (min) and operating temperature (°C) respectively.

Table III - Operating conditions using the SCCR5 for dextran fractionation

Run	Flow rates (cm^3/min)			Eluent to feed flow ratio	Switch time (min)	Feed conc % w/v	Operating temp °C	Cycle
	Feed	Eluent	Purge					
23.1-40-101.1-7.5-20	40	101.1	413	2.53	7.5	23.1	20	14
23.6-17.5-208.8-5-20	17.5	208.8	508	11.93	5	23.6	20	16
21.5-10.7-52.1-20-20	10.7	52.1	130	4.87	20	21.5	20	12
23.2-10.4-52.1-20-45	10.4	52.1	269	5.01	10	23.2	45	16

Table IV - Fractionation results obtained from the SCCR5 experiments

Run	HIGH MOL WEIGHT PRODUCT					LOW MOL WEIGHT PRODUCT				
	Conc %w/v	% of feed recovered	$\overline{M}_w$	$\overline{M}_n$	D	Conc % w/v	% of feed recovered	$\overline{M}_w$	$\overline{M}_n$	D
23.1-40-101.1-7.5-20	0.94	14.1	123600	39000	3.17	1.82	82.0	49300	23700	2.03
23.6-17.5-208.8-5-20	1.45	78.9	78700	29700	1.98	0.09	10.3	33100	19000	1.74
21.5-10.7-52.1-20-20	3.12	86.3	73400	30900	2.38	0.16	9.4	39300	20700	1.90
23.2-10.4-52.1-20-45	3.52	90.5	69300	25600	2.71	0.09	6.5	36200	16800	2.15

TableV- Removal of HMW from the dextran feed using the SCCR5 mk 2

Run	Flow rates (cm^3/min)			Eluent to feed flow ratio	Switch time (min)	Feed conc % w/v	LMW Product		
	Feed	Eluent	Purge				% of feed recovered	$\overline{M}_w$	$\overline{M}_n$
13.4-60-95-7.5-60	60	95	300	1.58	7.5	13.4	76.5	29000	7500
13.1-40-95-7.5-60	40	95	300	2.38	7.5	13.1	83.6	29000	8100
13.1-60-96-7.5-60*	60	96	300	1.6	7.5	13.1	85.1	32000	7500
7.1-25-116-7.5-60*	25	116	325	4.64	7.5	7.1	88.6	30150	7150
7.1-25-125-7.5-60*	25	125	325	5.00	7.5	7.1	74.6	27000	6000

* The columns were packed with Spherosil XOB030

Table VI - Dextran fractionation results obtained using the BCR1

				PRODUCT				
Run	Eluent flow rate (cm^3/min)	Feed charge (cm^3)	Feed conc % w/v	$\overline{M}_w$	$\overline{M}_n$	M_{10}	M_{90}	% of dextran in feed recovered
16.9-102-100-20	100	102	16.9	28100	16500	13000	68000	78.6
17.2-226-100-20	100	226	17.2	36300	20600	15000	126000	79.5
26.3-239-100-20	100	239	26.3	37800	21700	14000	123000	84.6
16-149-100-20	100	149	16.0	31200	17600	14000	119000	82.4
26.8-145-100-20	100	·145	26.8	36200	23000	15000	144000	83.3

Run code indicates feed concentration (% w/v), feed volume (cm^3), eluent flow rate (cm^3/min) and operating temperature (°C).

Table VII - Anion exchange chromatography - operating conditions

Run	Flow rates (cm^3/min)			Feed conc (% w/v)			Switch time (min)	Temp °C	Eluent to feed ratio
	Feed	Eluent	Purge	G	F	Total			
51.6-7-28-30-60	7	28	120	25.5	26.1	51.6	30	60	4.0
49.9-10-28-30-60	10	28	120	24.7	25.2	49.9	30	60	2.8
50.0-7-30-30-60	7	30	120	24.5	25.5	50.0	30	60	4.3
50-7-30-30-45	7	30	120	24.5	25.5	50.0	30	45	4.3
49.1-7-30-30-45	7	30	120	Inverted sucrose		49.1	30	45	4.3

Table VIII - Anion exchange chromatography results

	GLUCOSE RICH PRODUCT			FRUCTOSE RICH PRODUCT		
Run	% of glucose in feed recovered	Purity %	Bulk conc % w/v	% of fructose in feed recovered	Purity %	Bulk conc % w/v
51.6-7-28-30-60	95	67	1.7	51	99	2.8
49.9-10-28-30-60	95	75	1.9	59	84	3.4
50-7-30-30-60	93	84	1.6	80	97	3.8
50-7-30-30-45	94	89	1.9	83	97	3.9
49.1-7-30-30-45	95	88	1.8	84	96	4.1

Table IX - Operating conditions using the SCCR6 equipment

Run	Flow rates (cm^3/min)			Feed concentration (% w/v)				Operating temp °C	Switch time (min)	Eluent to feed ratio
	Feed	Eluent	Purge	Glucose	Fructose	Dextran	TOTAL			
20-35-105-30-20	35	105	550	10.0	10.0	-	20.0	20	30	3
40-35-105-30-20	35	105	550	20.0	20.0	-	40.0	20	30	3
50-35-105-30-20	35	105	550	25.0	25.0	-	50.0	20	30	3
69.4-70-210-15-60*	70	210	550	6.2	48.3	14.9	69.4	60	15	3

* The fructose rich product was used as eluent.

Table X - Results obtained using the SCCR6 equipment

Run	GLUCOSE/DEXTRAN RICH PRODUCT					FRUCTOSE RICH PRODUCT			
	Purity	Mass balance %	Product Conc % w/v			Purity %	Mass Balance %	Product conc. % w/v	Throughput sugar solids kg/h
			Glucose	Dextran	Total				
20-35-105-30-20	99.9	101	2.37	-	2.37	99.9	98	0.64	0.42
40-35-105-30-20	99.9	96	3.6	-	3.6	88.0	96	1.29	0.84
50-35-105-30-20	99.9	103	4.5	-	4.5	82.0	95	1.74	1.05
69.4-70-210-15-60*	72.0	101	9.7	4.2	12.9	99.9	114	16.3	2.94

* The fructose rich product was used as eluent.

Table XI - HFS production using the SCCR7 mk2 - Operating conditions

Run	FLOW RATES (cm³/min)			FEED CONCENTRATION % w/v				Switch time (min)	Temp. °C	Eluent to feed ratio	Fructose to glucose ratio in feed	Throughput (kg/h) sugar solids
	Feed	Eluent	Purge	Glucose	Fructose	Maltose +Oligos	TOTAL					
36-13-40-23-60	13	40	80	18.77	15.19	2.06	36.02	23	60	3.08	42.2/52.1	0.281
37-13-40-25-60	13	40	80	19.6	15.9	1.5	37	25	60	3.08	43/52.3	0.288
46-13-39-24.17-60	13	39	80	23.98	19.22	2.8	46	24.17	60	3.0	41.8/52.1	0.359
54-13-39-24.5-60	13	39	80	28.08	22.73	3.19	54	24.5	60	3.0	42.1/52	0.421
47.5-13-32-28.5-60	13	32	60	25.2	19.7	2.6	47.5	28.5	60	2.46	41.5/53	0.371
66-14.6-40-25-60*	14.6	40	70	34.3	27.8	3.9	66	25	60	2.74	42.1/52	0.578
66.3-14.6-40-26.5-60**	14.6	40	70	34.57	27.82	3.9	66.3	26.5	60	2.74	42/52.1	0.581

* In this run the product splitting technique was employed and the results correspond to the concentrated products only.

** In this run the product splitting technique was employed and the diluted splits were recycled as eluent and purge water.

Table XII - HFS production using the SCCR7 mk2 - Results

| Run | GLUCOSE RICH PRODUCT | | | | | FRUCTOSE RICH PRODUCT | | | | |
	Glucose purity %	% of glucose in feed recovered	Conc. % w/v	Impurities % Fructose	Impurities % Maltose + Oligos	Fructose purity %	% of fructose in feed recovered	Conc. % w/v	Impurities % Glucose	Impurities % Maltose + Oligos.
36-13-40-23-60	83.04	88.1	4.49	5.49	11.46	92.4	77.1	1.69	7.6	-
37-13-40-25-60	71.6	113.0	8.12	22.5	5.9	100	56.6	1.2	-	-
46-13-39-24.17-60	80.84	83.1	5.17	6.1	13.06	91.78	82.8	2.19	8.22	-
54-13-39-24.5-60	81.11	86.58	6.94	6.35	12.54	91.35	88.61	3.12	8.65	-
47.5-13-32-28.5-60	77.6	73.7	7.44	9.4	13.0	76.62	73.74	4.07	23.38	-
66-14.6-40-25-60*	84.83	94.23	22.56	4.49	10.67	94.8	95.78	11.29	5.2	-
66.3-14.6-40-26.5-60**	86.22	84.1	25.4	6.69	7.19	90.2	87.8	12.96	8.87	0.93

* In this run the product splitting technique was employed and the results correspond to the concentrated products only.

** In this run the product splitting technique was employed and the dilute splits were recycled as eluent and purge water.

Table XIII - Batch chromatographic separation results (BCR2)

Run	Feed Conc. %w/v	Feed Volume (cm³)	Through-put sugar Solids (kg/h)	GLUCOSE RICH PRODUCT			FRUCTOSE RICH PRODUCT			RECYCLE			
				% of total sugars in feed recovered	Glucose purity %	Conc. % w/v	% of total sugars in feed recovered	Fructose purity %	Conc. % w/v	% of total sugars in feed recycled	Composition % Fructose	Glucose	Total Conc. % w/v
20-15000-105-20	20	15000	0.056	39.7	99.9	7.65	32.8	99.9	5.44	27.5	54.2	45.8	12.62
40-15000-105-20	40	15000	0.257	47.9	99.9	15.52	37.2	99.0	10.43	14.9	72.6	27.4	20.69
60-15000-105-20	60	15000	0.495	31.9	99.9	21.85	28.8	99.3	15.15	39.3	71.2	28.8	33.70
20-1200-105-20	20	1200	1.065	21.2	99.9	0.98	24.5	99.9	3.57	54.3	82.3	17.7	4.07
20-6000-105-20	20	6000	1.500	38.6	99.9	5.27	29.8	99.9	5.44	31.6	54.2	45.8	12.62

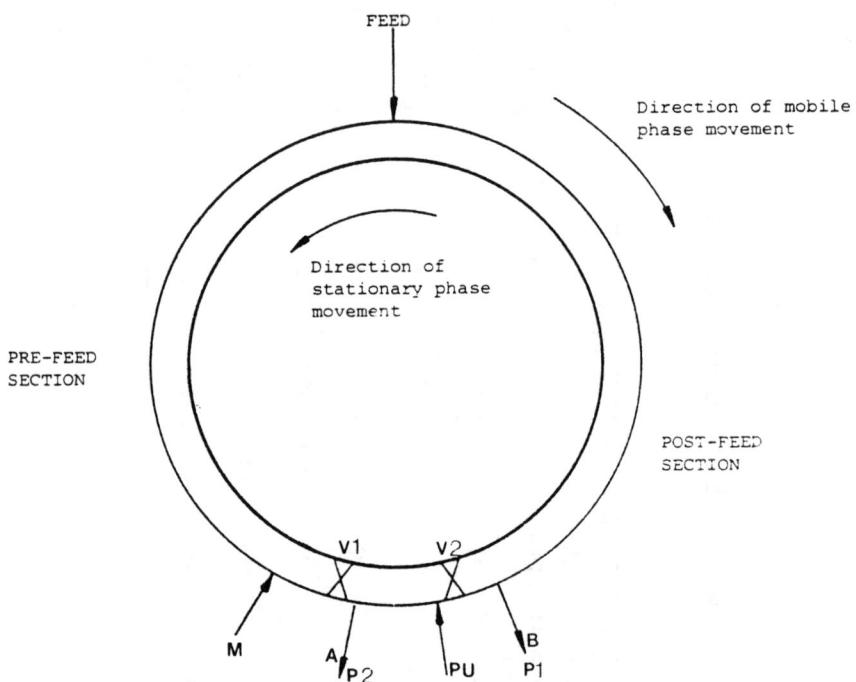

FIGURE 1:   Diagrammatic representation of the semicontinuous principle of operation

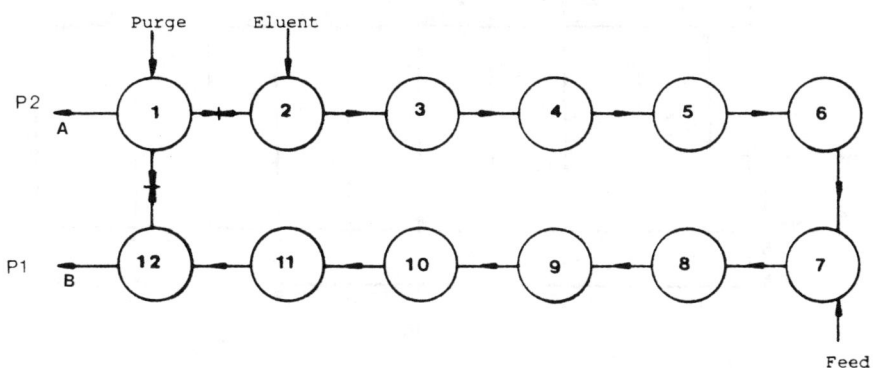

(a) SWITCH ONE

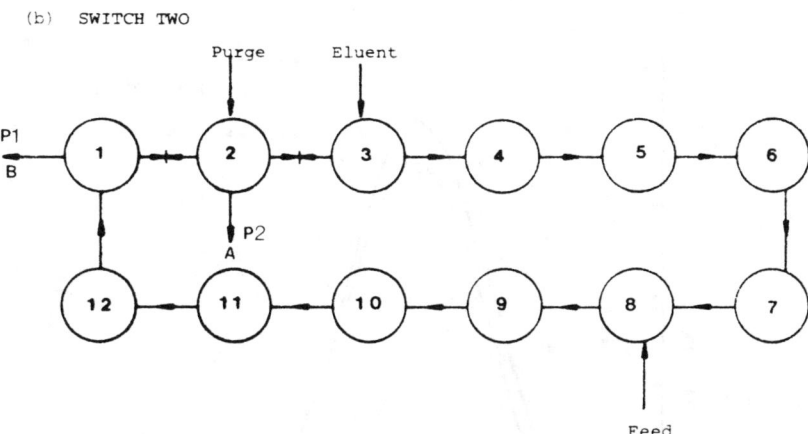

(b) SWITCH TWO

FIGURE 2: Sequential operation of the SCCR7 system

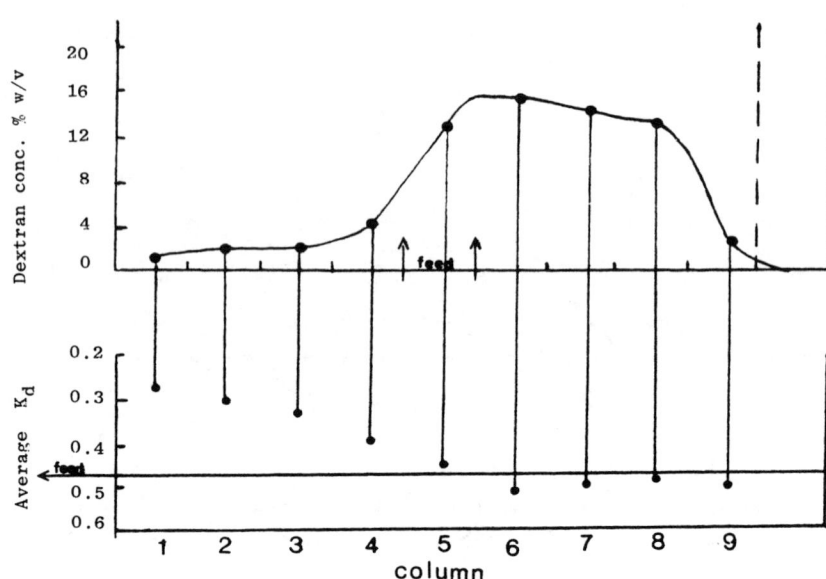

FIGURE 3:   Concentration and distribution coefficient profiles on the SCCR3 for a run 21– 7.9 – 43.9 – 17.6 – 20

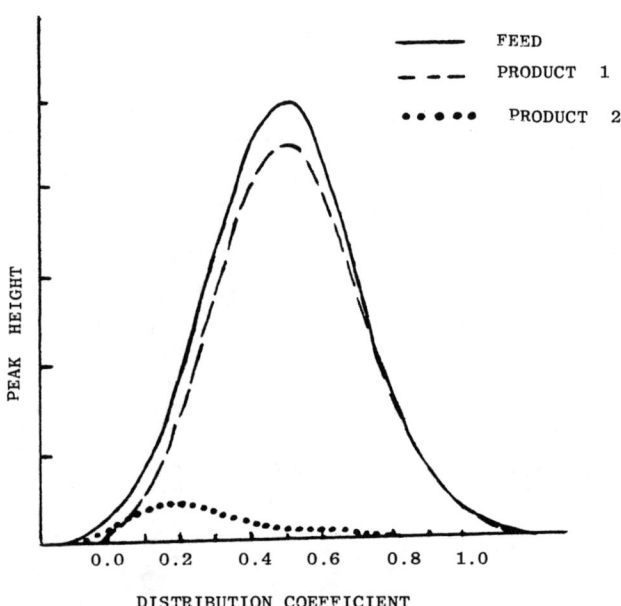

FIGURE 4:   Feed and product distributions for run 21 – 7.9 – 43.9 – 17.6 – 20

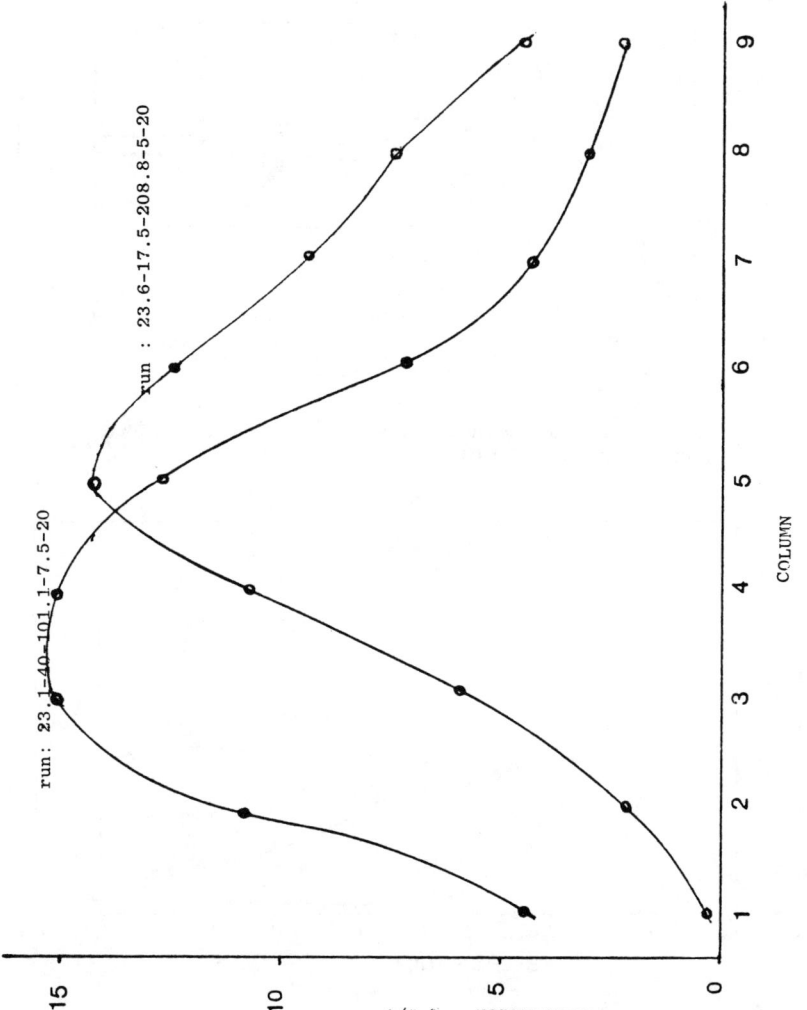

FIGURE 5: Dextran on column concentration profiles for two different runs

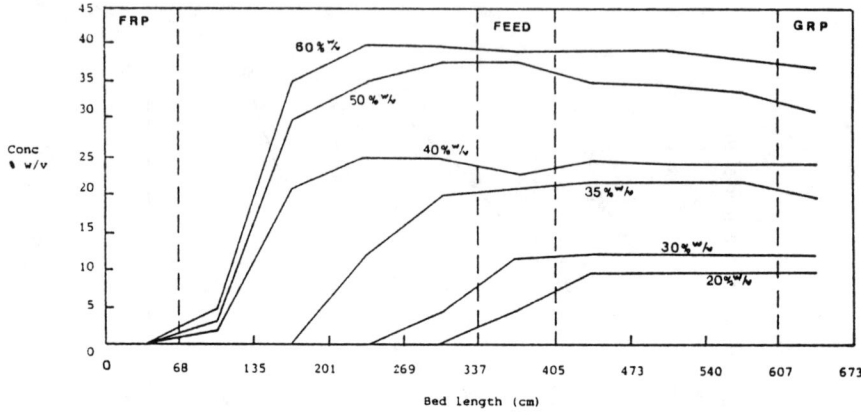

FIGURE 6:   Glucose on column concentration profiles at increasing feed
concentrations (SCCR6)

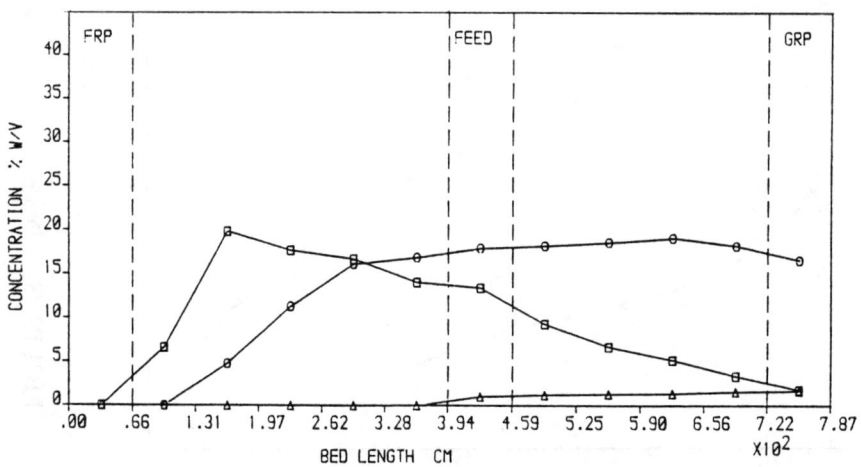

FIGURE 7:   On-column concentration profile for run 36 – 13 – 40 – 23 – 60
Key: □ Fructose   o Glucose   Δ Maltose + OS

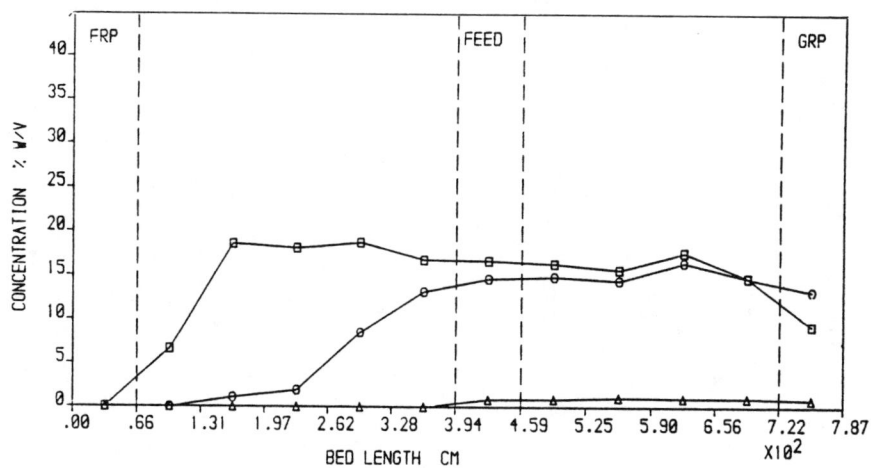

FIGURE 8: On-column concentration profile for run 37 – 13 – 40 – 25 – 60
Key □ Fructose o Glucose △ Maltose + OS

Chapter 10

Novel Applications of Continuous Annular Chromatography

Giorgio Carta
Charles H. Byers

Continuous Annular Chromatography (CAC) has been developed as a process that allows steady-state operation of multicomponent, chromatographic separations. The apparatus uses a slowly rotating annular bed of sorbent material. Feed is continuously introduced at a stationary point at the top of the annulus, while eluent flows over the rest of the circumference. The rotation of the sorbent bed coupled with down-flow of eluent and sorption, causes the separated components of the feed to form helical bands, which appear at the bottom of the annular bed at characteristic, stationary exit points. The process concept has been developed experimentally using columns ranging from 50 to 450 mm in diameter and operating between 7 and 1300 kPa. Three novel applications of CAC were investigated: the separation of sugars by simple elution, the separation of metal ion mixtures by multiple step elution, and the separation of amino acids by displacement development. Experimental results and theoretical developments in these fields of application are reviewed in this paper.

Chromatography has acquired in recent years a role of considerable importance in various industrial applications. The method has been broadly applied to mixtures of organic and inorganic substances. It is particularly useful with mixtures of compounds whose physicochemical properties are so close that other separation techniques are impractical. A great variety of sorbents is used to carry out chromatographic separations. However, although the chemistry and the efficacy of the various sorbent media differ considerably, the

† Operated by Martin Marietta Energy Systems, Inc., for the U.S. Department of Energy, under contract DE-AC05-840R21400.

underlying process concept is the same: a sample of the mixture of components that requires separation is applied to a chromatographic column and the components are separated along the column length as they migrate at different velocities under the effect of an eluent, which is continuously supplied to the column. The separated components are eventually recovered at the column exit end as they emerge at different times and the sample application-elution cycle is repeated.

Continuous operation is desirable in many industrial applications, and especially for large scale processes when the separation system may be coupled to other continuous unit operations. Continuous Annular Chromatography (CAC) has been developed as a broadly applicable process concept which allows truly continuous (*i.e.* steady-state) chromatographic separations. The CAC apparatus consists of an annular bed of adsorbent particles, packed in the space between two concentric cylinders, as illustrated schematically in Figure 1. While the column assembly is slowly rotated about its axis, eluent and feed solutions are continuously supplied to one end of the annular bed. To obtain an isocratic separation of the feed components, the eluent is uniformly fed to the entire bed circumference at the top of the annulus, while the feed mixture is introduced into a narrow sector at a point that remains stationary in space. As time progresses, stationary helical component bands develop from the feed point, with slopes dependent upon eluent velocity, rotational speed, and the distribution coefficient of the component between the fluid and adsorbent phases. At steady state the components form helical bands between the feed sector at the top of the bed and the bottom of the annular bed. Here the separated components are continuously recovered at various points along the circumference. As also pointed out by Wankat (1), CAC is a two-dimensional chromatographic process that replaces the length and time coordinates characteristic of conventional chromatography with the length and angular displacement coordinates of the rotating bed. In this respect, conventional chromatography and CAC are completely analogous and, in principle, CAC can perform continuously any separation that conventional chromatography can perform in a batch-wise fashion. In practice, of course, there may be some limitations associated with flow distribution, materials of construction in relation to their compatibility with the feed and eluent, high pressure operations, very large scale of operation, and three-dimensional dispersive effects in the annular bed.

Various attempts to develop continuous chromatographic systems have been made in the past and several conceptual designs have been proposed (2 - 7). Martin (2) is credited with the first proposal of an annular chromatograph for industrial applications. Theoretical considerations on the design of rotating chromatographs for gas separations were given by Giddings (3), while Taramasso and coworkers (4 - 6) reported the experimental development of a semicontinuous unit consisting of a carousel of 100 conventional fixed-bed columns. Arrangements of columns similar to that proposed by Taramasso and coworkers were also devised by Byalyi and Ganitskii (7), Witcherle and Coupek (8), and Dunnill and Lilly (9), for liquid chromatography applications. Fox et al. (10) and Moskvin and coworkers (11 - 13) reported the construction of a rotating chromatograph similar to that shown in Figure 1. Various applications, both in liquid and gas chromatography, were proposed and tested by these authors. A rotating chromatograph was also described by Cho et al. (14) to carry out simultaneously solid-catalyzed reactions and chromatographic separations on a continuous basis.

Since 1974 work has been in progress at Oak Ridge National Laboratory (ORNL) to investigate the feasibility of using CAC for preparative and industrial scale separations. This work was initiated with the development of a pressurized CAC apparatus capable of operating at throughputs much higher then previous devices, which were limited to gravity flow (15). Following this initial development, several CAC units with diameter ranging from 89 to 450 mm, annulus width from 6.4 to 51 mm, and bed length from 0.6 to 1.1 m, were constructed and operated at ORNL (16 - 18). A detailed review of the construction details of the various units is given by Begovich (19).

A few applications of CAC have been investigated systematically, including the separation of Ni, Cu, and Co in ammoniacal solutions (19), the separation of iron and aluminum in ammonium sulfate-sulfuric acid solutions (20), and the separation of hafnium from zirconium in sulfuric solutions (20). The bulk of the experimental work at ORNL has been carried out with Dowex 50W-X8 resin as the sorbent. The resin is a polystyrene-divinylbenzene, strong-acid cation exchange resin with 8% degree of cross-linking. Size exclusion chromatography, using a mixture of Blue Dextran and $CoCl_2$ as a model systems was also investigated using Sephadex G15 as the separation medium (21).

More recent work on the CAC has been conducted through a collaborative research effort at ORNL and at the University of Virginia (22 - 26). This work has focussed on: (1) novel applications of CAC in the food, hydrometallurgical, and biotechnology fields, and (2) improved modes of operation of CAC, such as step elution and displacement development. The recent progress in these areas is reviewed in this paper. The emphasis is on three specific experimental systems: the separation of sugar isomer mixtures, the separation of iron-chromium mixtures, and the separation of amino acid mixtures. The first application is an example of an industrially important large scale fractionation which is carried out by isocratic elution chromatography. The second application is a model system for hydrometallurgical separations which demonstrates the use of step elution. Finally, the third application is an industrially important biochemical separation which illustrates the use of displacement development to obtain simultaneous separation and concentration.

Equipment

Figure 2 illustrates the construction details of the 279-mm-diameter CAC unit used in this study. The sorbent (Dowex 50W-X8 resin, with a wet particle diameter of 50-60 μm) is packed in the 1.27-cm-wide annulus formed between two concentric cylinders. The two cylinders are flanged at the bottom and constructed from clear Plexiglas. The inner cylinder is 80 mm shorter than the outer one and is closed at the top, providing head room for stationary feed and eluent lines.

A layer of spherical glass beads, 0.18 mm in diameter, is placed on top of the resin bed to facilitate feed and eluent distribution. Eluents and feed are introduced through stationary stainless steel nozzles, whose tips are located within the layer of glass beads. As the bed rotates, the nozzles make headway through the glass beads which are sufficiently fluid to prevent the formation of a permanent trench. One of the eluents may also be allowed to fill the head space above the layer of glass beads with little or no mixing with the other eluents which are injected within the layer of glass beads. The head space is sealed to a central shaft with a Teflon O-ring. The effluent from the annular bed is collected at the bottom through 180 eluate exit tubes. Porous polyethylene plugs are fitted in the exit holes and provide support for the resin bed. A digitally controlled drive system is used to slowly rotate the entire assembly while the feed and eluents distributor remains stationary. Feed and eluents are supplied to the CAC by constant-flow, positive displacement pumps.

The exit concentration profile at the bottom of the annular bed can be readily determined by connecting an in-line, continuous detection instrument with a single eluate exit tube. In this manner, as the CAC unit rotates the detector receives eluate from all circumferential points as the apparatus completes a 360 degree turn. Thus, if the concentration-detection instrument is attached to a recording device, an apparently conventional (*i.e.* concentration *vs* time) chromatogram is recorded. Angular concentration profiles are immediately obtained, considering that the angular displacement from the feed point, θ, is related to the recorded time, t, and to the rotation rate, ω, by $\theta = \omega t$. The connection between the eluate exit tube and the detector must be provided by a metering pump to compensate for the pressure drop through the instrument and ensure constant flow through the sample tube, equal to 1/180 of the total eluent flow. Naturally, liquid samples can also be collected directly at various angular locations, as it would be done in practice to recover the separated products. A more accurate determination of the angular concentration profiles is however possible with the former technique.

A number of variables may be independently manipulated to optimize the separation performance. These include feed flow rate and concentration, total flow rate, the relative flow rates of the different eluents, and the rotation rate.

Theory of Operation

Considering a uniformly-packed annular sorbent bed with void fraction ϵ, the following steady-state continuity equation may be written for a solute with concentration c

$$\epsilon D_z \frac{\partial^2 c}{\partial z^2} + \frac{\epsilon D_\theta}{R_0^2} \frac{\partial^2 c}{\partial \theta^2} = \omega \epsilon \frac{\partial c}{\partial \theta} + \omega(1 - \epsilon) \frac{\partial q}{\partial \theta} + u \frac{\partial c}{\partial z}, \tag{1}$$

where D_z and D_θ are the axial and angular dispersion coefficients, ω is the rate of rotation, R_0 is the radius of the annular chromatograph, u is the fluid superficial velocity, and z and θ the axial and angular coordinates respectively.

The concentration of solute in the sorbent, q, and the fluid phase concentration, c, are related by a rate equation. Using a film model to describe fluid-particle mass transfer, the following equation may be written

$$\omega(1 - \epsilon)\frac{\partial q}{\partial \theta} = k_o a \, (c - c^*), \tag{2}$$

where $k_0 a$ is an overall mass transfer coefficient and c^* is the fluid concentration in equilibrium with the solid.

If the angular dispersion term in Equation 1 is neglected and the transformation $\theta = \omega t$ is made, Equations 1 and 2 become

$$\epsilon D_z \frac{\partial^2 c}{\partial z^2} = \epsilon \frac{\partial c}{\partial t} + (1 - \epsilon)\frac{\partial q}{\partial t} + u\frac{\partial c}{\partial z}, \tag{3}$$

$$(1 - \epsilon)\frac{\partial q}{\partial t} = k_o a \, (c - c^*). \tag{4}$$

These equations correspond exactly the solute continuity equations describing unsteady operation of a fixed bed. Thus, as pointed out by Wankat (1), under conditions of negligible angular dispersion, the unsteady-state, one-dimensional chromatographic process is equivalent to the steady-state, two-dimensional CAC process. Solutions of Equations 3 and 4 available in the literature may be directly applied to predict the steady state performance of CAC operations. For systems exhibiting a linear adsorption isotherm various analytical solutions are available to describe isocratic elution.

Howard et al. (23) have demonstrated that at the flow rates typically encountered in liquid chromatography applications, when fluid and intraparticle mass transfer resistances are present, both angular and axial dispersion are only moderately important in the CAC. In this case, for an infinitesimally small feed pulse, the solute concentration profile is described by (27)

$$c(z, \hat{t}) = \frac{Q}{2\sqrt{\pi}} \left\{ \frac{(k_o a)^2}{u^3 z \hat{t} \, [(1 - \epsilon)K]^3} \right\}^{0.25} \exp\left\{ -\left[\sqrt{\frac{k_o a z}{u}} - \sqrt{\frac{k_o a \hat{t}}{K(1 - \epsilon)}} \right]^2 \right\}, \tag{5}$$

where K is the distribution coefficient for the solute, and

$$\hat{t} = \frac{\theta}{\omega} - \frac{\epsilon z}{u}. \tag{6}$$

Equation 5 applies as an approximation when the number of transfer units available in the bed is greater than 5. The amount of solute introduced per unit cross sectional area of the sorbent bed, Q, may be calculated from

$$Q = \frac{c_F u Q_F}{Q_T} \frac{360°}{\omega},$$

(7)

where c_F is the solute concentration in the feed mixture, Q_F is the feed flow rate, and Q_T is the total flow rate of fluid through the annular bed.

Equation 5 predicts that the resolution of two species is independent of loading. This is a good approximation for low bed loadings, but is obviously unrealistic at high loadings. When the feed occupies a significant sector of circumference, the following series solution may be used to compute concentration profiles (28)

$$c(z,t) = \frac{c_F \theta_F}{\theta_F + \theta_E} + \frac{2}{\pi} \sum_{j=1}^{\infty} \left\{ \frac{1}{j} \exp\left[-\frac{j^2 k_o a z}{(j^2 + r^2)u} \right] \times \sin\left[\frac{j\pi\theta_F}{\theta_F + \theta_E} \right] \right.$$
$$\left. \times \cos\left[\frac{2j\pi\theta}{\theta_F + \theta_E} - \frac{j\pi\theta_F}{\theta_F + \theta_E} - \frac{2j\pi z\omega\epsilon}{u(\theta_F + \theta_E)} - \frac{jrk_o a z}{(j^2 + r^2)u} \right] \right\},$$

(8)

where θ_F is the arc over which feed is applied to the column, θ_E is the elution arc, and r is given by

$$r = \frac{k_o a(\theta_F + \theta_E)}{2\pi(1 - \epsilon)K\omega},$$

(9)

For systems which exhibit non-linear isotherms a numerical solution of Equations 3 and 4 is in general required to predict the performance of chromatographic separations (29).

Applications

Separation of Sugars -- Isocratic Operation

The separation of sugars, and especially the fractionation of glucose-fructose mixtures, is an important industrial problem. Typical fructose syrup, in fact, contains 42% fructose, 52% glucose, and 6% oligosaccharides on a dry-weight basis. Because of the lower cost of fructose syrup for equivalent sweetness fructose syrup is used as a sucrose replacement in some foods and beverages. Since fructose is considerably sweeter than glucose and is more soluble in water, syrup with a high fructose content is desirable for many applications. Because of the large scale of the operation, efficient adsorbent utilization and continuous operation are desirable. In the 'Sorbex' process, which is frequently used for this application (30 - 34), an effective countercurrent operation is achieved by the sequential periodic movement of feed and exit ports through a series of fixed beds. The adsorbent is used very efficiently in this process. On the other hand, however, the process is mechanically

complex and may not be suitable for all scales of operations. Furthermore, it can completely separate only two components, with all
other components accumulating in one of the two products.

The separation of mixtures of glucose, fructose, sucrose, and
higher molecular weight saccharides by CAC has been investigated
using the calcium form of Dowex 50W-X8 resin. The separation results in this case from specific interactions of the sugars with the
calcium counter-ions held by the resin, as well as from size exclusion effects (35 - 36). β-D-fructose is the most strongly adsorbed
species, followed by α- and β-D-glucose. While the β anomer is the
largely predominant form of fructose in aqueous solution, the two
anomeric forms of glucose coexist as a slowly equilibrating mixture
(37). Sucrose and other higher-molecular-weight oligosaccharides
interact only very weakly with the hydration sphere of the calcium
counter- ions, and they are partially or totally excluded from the
resin matrix owing to their larger molecular size.

The CAC was operated isocratically using distilled water containing 0.5 g/L of calcium chloride as the eluent the. The resin was
packed to a depth of 58 cm for these experiments. Fixed-bed runs were
also carried out to determine equilibrium and mass transfer parameters for the various species. A summary of the parameters determined
from these experiments is given in Table I.

Table I. Equilibrium Distribution and Mass Transfer Coefficients
Dowex 50W-X8, Calcium-form, T=25°C

Sugar	Distribution Coeff., K	Mass Transfer Coeff., $k_o a(s^{-1})$
sucrose	0.130	0.0042
β-D-glucose	0.228	0.0143
α-D-glucose	0.294	0.0184
β-D-fructose	0.657	0.0409

Figure 3 shows a typical CAC chromatogram for the separation of
a dilute glucose-fructose mixture containing 0.5 g/l of Blue Dextran
(M.W. 2,000,000), which was used as a tracer. Calculations based on
Equations 5 - 7 are also shown in this figure and are seen to be in
fairly good agreement with the experimental results. Since the model
parameter were derived from fixed-bed experiments, this provides
a good comparison of performance between fixed-bed and CAC operation. The effects of a number of parameters were investigated, as
described by Howard et al. (22 - 23). The effects of rotation rate
on the resolution are particularly important, since this parameter
determines the achievable production rate. Figure 4 shows experimental and calculated results using a feed rate of 1 cm³/min and an
eluent flow rate of 4.0 L/h. At a high rotation rates, the resin bed
loading is small and the resolution between glucose and fructose becomes independent of ω. At lower rotation rates, however, the resin
bed loading and productivity increase (see Equation 7), but the res-

olution is reduced. As shown by Figure 4, the experimental resolution values fall somewhat short of prediction based on fixed bed data. The differences may be attributed in part to somewhat greater dispersion in the CAC and to other operational difference as discussed by Howard (22).

The effects of feed concentration are also important since rather concentrated solutions would be used in practice. As has been observed in previous studies on the sugar separation system (31 - 34), no significant effect of concentration on elution time or angle was noted for sugar concentrations up to 200 g/L both in fixed-bed and in CAC operation (24).

The resolution was however reduced significantly with more concentrated solutions, largely because of the non-uniformity of flow through the annular bed caused by the increased viscosity of the feed relative to the viscosity of the eluent. Operation at higher temperature (60 °C is customary for the glucose/fructose fractionation) would mitigate the flow non-uniformity problem with only a small effect on the distribution coefficients (36). Pilot scale experiments were also conducted for various sugar mixtures (24). A stainless steel CAC unit 44.5 cm in diameter and with an annulus thickness of 3.2 cm was employed for this work. The unit was packed to a depth of 107 cm with Dowex 50W-X8 in the calcium form. Successful separations, reported by Byers et al. (24), were obtained with industrial mixtures derived from corn syrup and inverted beet sugar molasses.

Separation of Metal Ions -- Step Elution

Hydrometallurgical separations in which various metals are simultaneously recovered are often encountered in the metal- process industries. Solvent extraction generally allows only the recovery of one of the species, while the remaining metal values are left as a mixture in the raffinate phase. Where it desirable to recover several metals in pure form, several processing steps are necessary, and few economic options are available. CAC may provide a viable alternative to conventional processing techniques.

The chemical system selected to explore the application of CAC is a typical metal separation of the type required in the processing of low-grade ores, such as in the dump or heap leaching of ores (38). For the sake of simplicity in continuously analyzing the mixtures, iron and chromium were chosen as model components. The feed consisted of ferric and chromic nitrate salts (5 g/L, of each metal) dissolved in distilled water. The eluents were acid ammonium sulfate solutions varying in concentration between 0.4 M and 1.5 M, with a constant 0.025 M concentration of sulfuric acid. Dowex 50W-X8 resin packed to a depth of 40 cm was again used for the experimental work which is described in detail by Byers et al. (25).

The separation was carried out by CAC both in isocratic and step elution modes. In the latter mode of operation, two or more eluents of increasing strength are used, allowing an enhancement of the separation by hastening the elution of strongly retained components. This results into two distinct beneficial effects: (1) the productivity is increased since the separation will occur over a narrower sector; and (2) the peaks will be sharpened and product dilution will be reduced. The distribution of eluents along the bed circumference in a typical multiple step elution CAC experiment is shown in Figure 5. In the case illustrated two eluents are used: the main eluent, a 0.4 M $(NH_4)_2SO_4$ solution, which immediately follows the feed; and a second eluent, a 1.0 M $(NH_4)_2SO_4$ solution, which is applied at a position 20 degrees from the feed point, and covering a 40 degree sector. Visual observation of the component bands revealed a very sharp boundary between the two eluents. As shown by Byers et al. (25), the distribution coefficients for Fe^{3+} and Cr^{3+} ions decrease exponentially with the concentration of ammonium sulfate. Thus, faster elution and band sharpening of the two species may be expected as the $(NH_4)_2SO_4$ concentration is increased.

A comparison of exit concentration profiles for an isocratic run and for a step elution run is given in Figure 6. The two experiments were carried out for the same conditions, except for the latter one involved a step in the eluent concentration over the annulus. It is apparent that, while the bed height was not optimized for the isocratic separation, the elution angle of the slowly eluting component (Cr^{3+}) is drastically reduced. Further, both peaks are considerably sharpened resulting in a greatly mitigated dilution effect. As shown by Byers and coworkers, with a judicious choice of process conditions, product dilution may be completely eliminated by step elution.

Separation of Amino Acids -- Displacement Development

A promising filed of application of CAC is that of bioseparations. Recent advances in biology need to be accompanied by advancements in the area of bioprocessing and development of effective and efficient separation techniques suitable for production scale applications is especially needed, since the impact of these operations is frequently dominant. While it is apparent that conventional chromatography is a very powerful tool in this area, it is also apparent that many systems may be more efficiently handled by continuous processing.

We have investigated the separation of mixtures of amino acids by CAC (26), as an example of a biochemical separation. Mixtures of L-leucine (Leu), L-valine (Val), and L-glutamic acid (Glu) dissolved in water have been used as a model system. As is discussed by Carta et al. (39), uptake of amino acids by the hydrogen form of cation exchange resins occurs primarily by means of the stoichiometric exchange of amino acid cations and hydrogen ions. Thus the ionization of the amino acid molecules in solution determines the amount of amino acid taken up as a function of solution pH and concentration. The separation of Glu from Val using Dowex 50W-X8 resin is based largely upon differences in the isoelectric pH values of the two amino acids, which are about 3.2 for Glu and about 5.9 for Val. Conversely, since the isoelectric pH values of Val and Leu are very close, the separation of these two amino acids is largely dependent upon differences in the specific affinities of the two amino acid cations for the resin.

The CAC was operated in displacement development mode using a packed-bed height of 20 cm. Figure 7 illustrates the arrangement of feed and eluents for this operation. The feed consists of a dilute aqueous solution (15 mM) of the three amino acids. A 0.1 M NaOH solution is used as the displacer. At pH-values above the isoelectric pH, in fact, the amino acids are almost entirely in the negatively charged form which is excluded from the resin by the Donnan potential effect. A 0.1 N sulfuric acid solution is used to reconvert the resin to the hydrogen form, and distilled water is used to remove the excess acid from the intraparticle bed voids. All four solutions flow simultaneously through the annulus. In this mode of operation, separation of the feed constituents can be obtained simultaneously to their concentration.

The achievable concentration level, which is obtained when the sorbent bed is sufficiently long to reach isotachic conditions, may be determined readily from the pure component uptake isotherms using the method developed by Glueckauf (40). The equilibrium uptake of Glu, Val, and Leu by the Dowex resin has been investigated by Carta et al. (39), and the pure component uptake isotherms are illustrated in Figure 8. Since the resin's functional groups are strongly acidic, NaOH is taken up with an essentially rectangular isotherm. To determine the limiting concentration level, one may simply draw a straight line between the origin and the point on the displacer isotherm corresponding to the displacer feed concentration. The isotachic concentrations are found at the intercepts of this line with the pure component isotherms. The separated species will elute in the order of increasing isotachic concentration. From the graphical construction it is apparent that the isotachic concentrations depend only upon the displacer concentration, and not upon the concentrations of the components in the feed. For the conditions shown in Figure 8, although Glu and Val have approximately the same isotachic concentrations, Glu would exit first, because of its low isoelectric pH.

The exit concentration profiles for the separation of a 15 mM mixture of Glu, Val, and Leu are shown in Figure 9. For this run the feed occupies a sector of 45 degrees, while the NaOH displacer and the acid regenerant occupy 90 degrees each. The rest of the annulus is occupied by distilled water which fills the overhead compartment. The total flow rate though the bed was 400 cm^3/min. As is apparent from these experimental results, while it is obvious that isotachic conditions were not perfectly reached in this run, the concentration profiles approach ideal conditions, demonstrating that separation and concentration can be obtained simultaneously and continuously by CAC. In this run the separated components are recovered at an average concentration of about four times that of the feed, with a product purity of 90%. Figure 9 also illustrates how the separation can be observed by following the pH profiles. Since each amino acid tends to buffer near its isoelectric point, the pH, which is initially close to 6, decreases to about 3.2 as Glu is eluted. The pH then increases gradually toward the isoelectric pH values of Val and Leu, and then very rapidly toward the pH of the NaOH displacer solution used. Carta and coworkers (39) describe mathematical approaches for modeling the operation.

Concluding Remarks

We have illustrated three novel applications of CAC in the food, hydrometallurgical, and biotechnology areas. For all applications examined the CAC allows truly continuous separation with a separation performance which, in most cases, is close to that of an equivalent, batch-wise chromatographic system.

The CAC separations can be carried out isocratically, by step elution, and by displacement development. High resolution isocratic separations can be obtained with a very simple arrangement involving a simple eluent, at the expense, however, of significant dilution of the products. With step elution, removal of slowly eluting components can be hastened increasing the productivity of the chromatograph. In addition, the dilution effect may be greatly mitigated or even completely eliminated, as the peaks become sharpened. Further sharpening of the product peaks could be obtained by arranging a series of eluent concentrationj steps along the annulus to simulate gradient elution operation, as is discussed by Byers et al. (25). Finally, in the displacement development mode of operation, separation and concentration are obtained simultaneously. Here four fluid streams generally have to flow simultaneously through the annulus. In this mode of operation, if the sorbent bed is sufficiently long, the product concentration depends only upon the pure component isotherms and the displacer concentration, and concentration of dilute solutions may be accomplished.

In conclusion, the CAC appears to be suitable for a variety of separation tasks. The apparatus allows continuous operation while retaining the desirable characteristics of multicomponent resolution capability and flexibility which are typical of conventional chromatography. Mechanical implementation of the process concept is also rather simple, at least at low to medium pressures of operation, and no valves are required. Scale-up of this process which is promising for many applications has also been demonstrated with a 18-in-diameter pilot scale CAC unit in operation at ORNL.

Acknowledgments

This research was sponsored, in part, by the Office of Industrial Programs, U.S. Department of Energy under contract DE-AC05-84OR21400 with Martin Marietta Energy Systems, Inc.. The experimental participation of Warren G. Sisson is greatly appreciated.

Legend of Symbols

a	=	particle-fluid interfacial area
c	=	liquid-phase solute concentration
c_F	=	feed solute concentration
d_p	=	particle diameter
D_z	=	axial dispersion coefficient
D_θ	=	angular dispersion coefficient

k_o = global mass transfer coefficient
K = equilibrium distribution coefficient
q = sorbent solute concentration
Q = column loading of solute
Q_F = feed flow rate
Q_T = total flow rate
R_0 = radius of annular bed
t = time
$\hat{t}$ = chromatographic time
u = superficial velocity
z = bed axial position
ϵ = bed void fraction
θ = displacement from feed point
θ_E = elution angle
θ_F = feed angle
ω = rotation rate

Literature Cited

1. Wankat, P. C. AIChE J. 1977, 23 859.
2. Martin, A. J. P. Discuss. Faraday Soc. 1949, 7, 332.
3. Giddings, J. C. Anal. Chem. 1962, 34, 37.
4. Dinelli, D.; Polezzo, S.; Taramasso, M. J. Chromatogr. 1962, 7, 477.
5. Polezzo, S.; Taramasso, M. J. Chromatogr. 1963, 11, 19.
6. Taramasso, M. J. Chromatogr. 1970, 49, 27.
7. Byalyi, A.L.; Ganitskii, M.B. USSR Patent 257,134, 1969.
8. Witcherle, O.; Coupek, J. Czech. Patent 152,620, 1974.
9. Dunnill, P.; Lilly, M.D. In Biotechnology and Bioengineering Symposium, No. 3, Gaden, E.L., Ed.; Wiley: New York, 1972, p. 97.
10. Fox, J.B.; Calhoun, R.C.; Eglington, W.J.; Nicholas, R.A. J. Chromatogr. 1969, 43, 48.
11. Moskvin, L.N.; Kozhin, S.A.; Fleisher, A.Y. J. Appl. Chem. USSR. 1974, 44, 2056.
12. Moskvin, L.N.; Gumerov, M.F.; Gorshkov, A.I. J. Appl. Chem. USSR. 1974, 47, 1917.
13. Moskvin, L.N.; Mozzhuklin, A.V; Tsaritsyna, L.G. J. Anal. Chem. USS 1975, 30, 2056.
14. Cho, B.K.; Carr, R.W.; Aris, R. Sep. Sci. Technol. 1980, 15, 679.
15. Scott, C. D.; Spence, R. D.; Sisson, W. G. J. Chromatogr. 1976, 126, 381.
16. Canon, R. M.; Sisson, W. G. J. Liq. Chromatogr. 1978, 1, 427.
17. Begovich, J. M.; Byers, C. H.; Sisson, W.G. Sep. Sci. Technol. 1983, 18, 1167.
18. Byers, C. H. MIT School of Chemical Engineering Practice; 1980-81 Report, Cambridge, 1982.
19. Begovich, J. M. Ph.D. dissertation, The University of Tennessee, Knoxville, 1982.
20. Canon, R. M.; Begovich, J. M.; Sisson, W. G. Sep. Sci. Technol.

1980, <u>15</u>, 655.
21. Sisson, W.G.; Begovich, J.M.; Byers, C.H.; Scott, C.D. Paper presented at <u>ACS National Meeting</u>, New Orleans, 1987.
22. Howard, A.J. M.S. Thesis, University of Virginia, Charlottesville, 1987.
23. Howard, A.J.; Carta, G.; Byers, C.H. ''Separation of Sugars by Continuous Annular Chromatography", Oak Ridge National Laboratory Report ORNL/TM-10318, 1987.
24. Byers, C.H.; Sisson, W.G.; Carta, G. Paper presented at <u>Tenth Symposium on Biotechnology for Fuels and Chemicals</u>, Gatlinburg, 1988.
25. Byers, C.H.; DeCarli, J.P.; Carta, G. Paper presented at <u>ACS National Meeting</u>, Toronto, 1988.
26. DeCarli, J.P. PhD Dissertation, University of Virginia, Charlottesville, 1988.
27. Sherwood, R. K.; Pigford, R. L.; Wilke, C. R. <u>Mass Transfer</u>; McGraw-Hill: New York, 1975.
28. Carta, G. <u>Chem. Eng. Sci.</u> 1988, <u>43</u>, 2877.
29. Ruthven, D. M. <u>Principles of Adsorption and Adsorption Processes</u>; John Wiley and Sons: New York, 1984.
30. Broughton, D. B. U.S. patent 3,291,726.13, 1966.
31. Ching, C. B.; Ruthven, D. M. <u>Can. J. Chem. Eng.</u> 1984, <u>62</u>, 398.
32. Ching, C. B.; Ruthven, D. M.; Hidajat, K. <u>Chem. Eng. Sci.</u> 1985, <u>40</u>, 1411.
33. Barker, P. E.; Thawait, S. <u>J. Chromatogr.</u> 1984, <u>295</u>, 479.
34. Barker, P. E.; Thawait, S. <u>Chem. Eng. Res. and Dev.</u> 1986, <u>64</u>, 302.
35. Goulding, R. W. <u>J. Chromatogr.</u> 1975, <u>103</u>, 229.
36. Welstein, H.; Sauer, C. In <u>Ion Exchange Technology</u>; Naden, D., Streat, M., Eds.; Ellis Harwood: London, 1984.
37. Lehninger, A. L. <u>Biochemistry</u>; Worth Publishers, Inc.: New York, 1977.
38. Wadsworth, M.E. <u>Sep. Sci. Technol.</u> 1987, <u>22</u>, 711.
39. Carta, G.; Saunders, M.S.; DeCarli, J.P.; Vierow, J.B. Paper presented at <u>AIChE Annual Meeting</u>, New York, NY, November, 1987.
40. Glueckauf, E. <u>Trans. Farady Soc.</u> 1955, <u>51</u>, 1540.

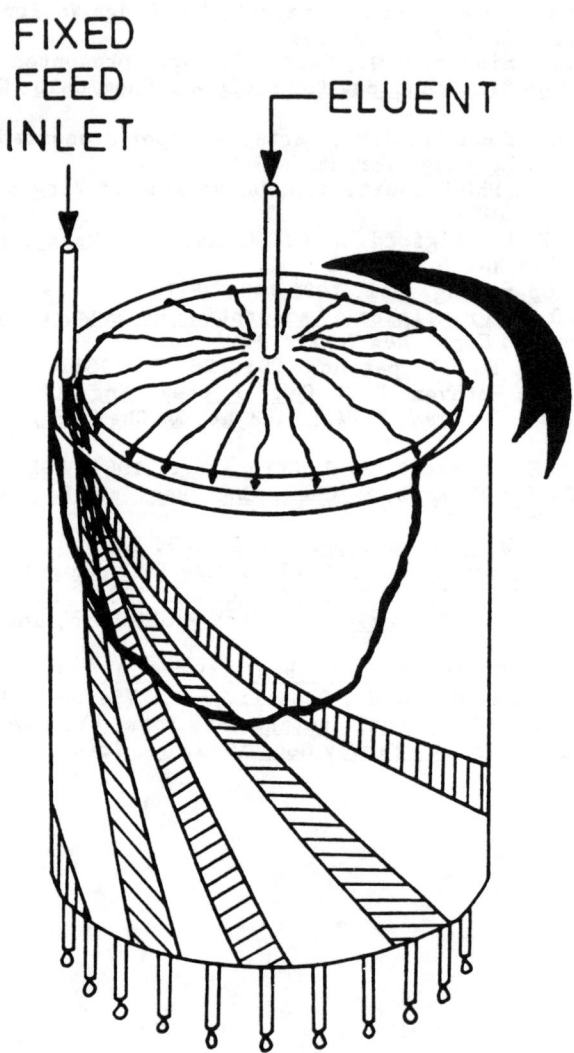

Figure 1. Conceptual view of the CAC apparatus.

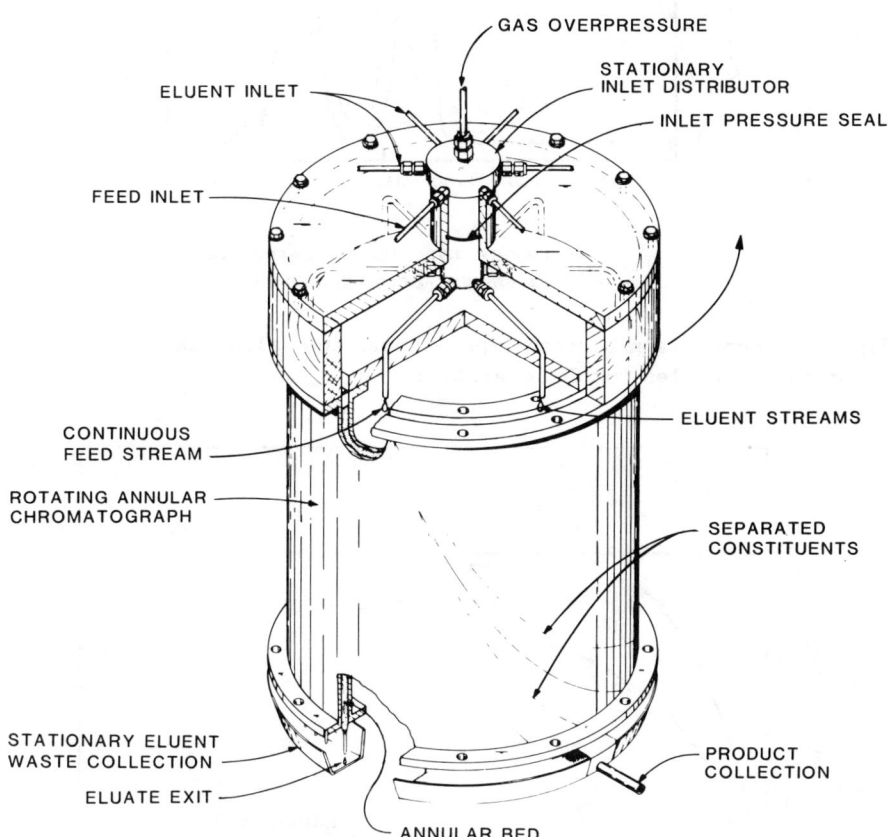

Figure 2. Construction details of the CAC apparatus.

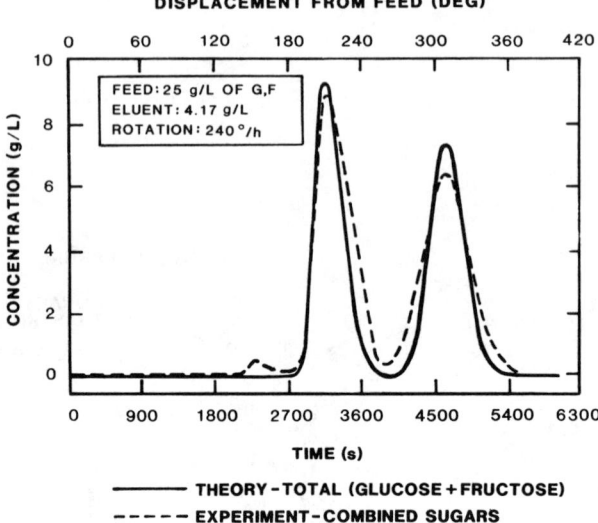

Figure 3. Comparison between experimental and calculated concentration profiles for CAC operation.

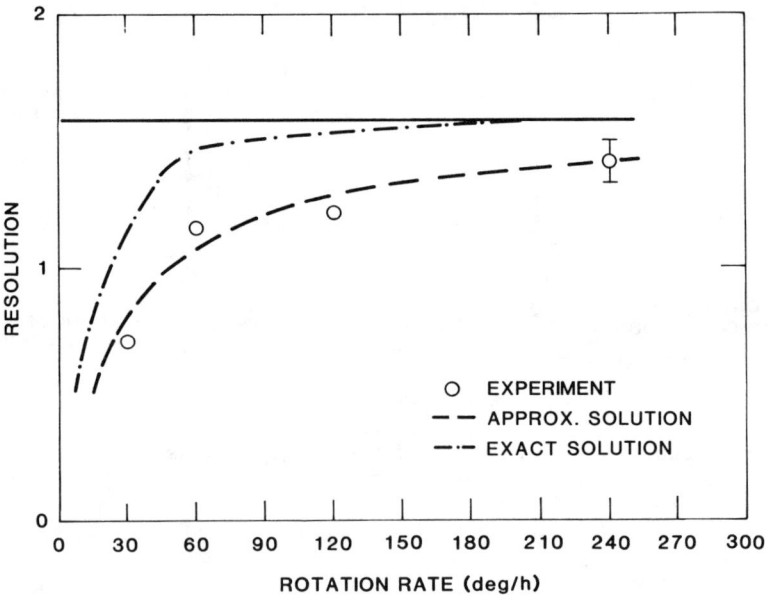

Figure 4. Effect of rotation rate on glucose-fructose resolution, Q_F = 1 mL/min, Q_T = 4.0 L/h. Feed contains 25 g/L glucose, 25 g/L fructose, and 0.5 g/L dextran.

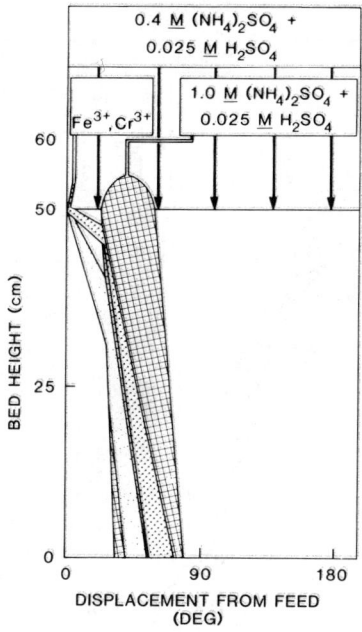

Figure 5. Step elution chromatogram, showing the motion of the two feed components and eluents as a function of angular position.

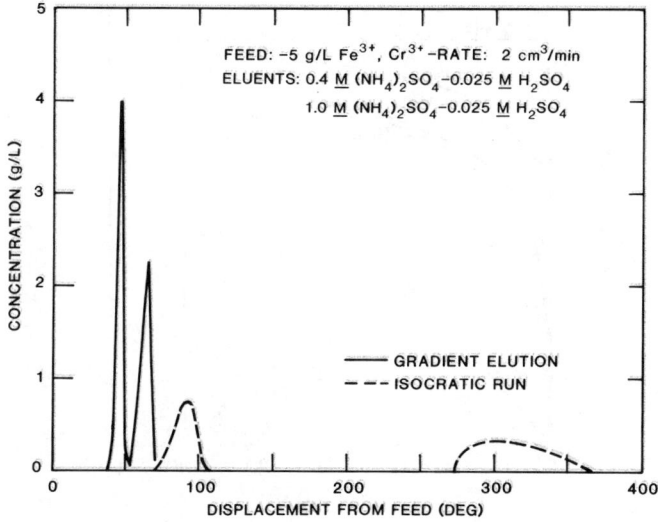

Figure 6. Separation of an iron-chromium mixture by isocratic and step elution by CAC.

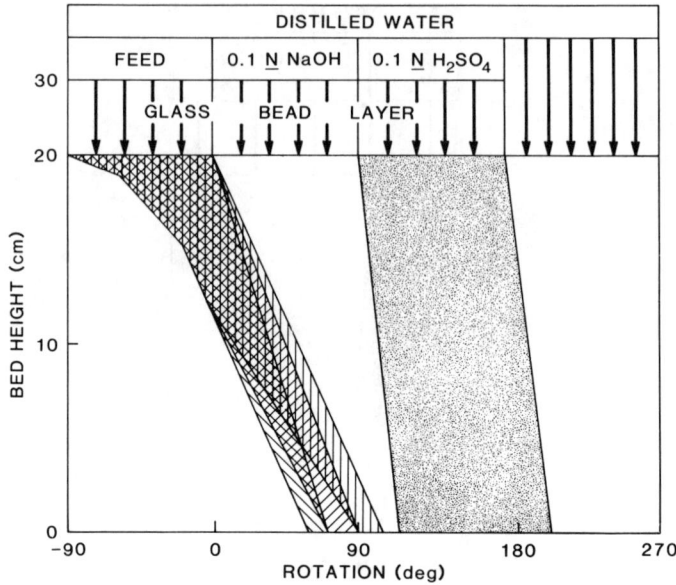

Figure 7. Arrangement of feed and eluents in displacement development showing band position as a function of bed height and angular position.

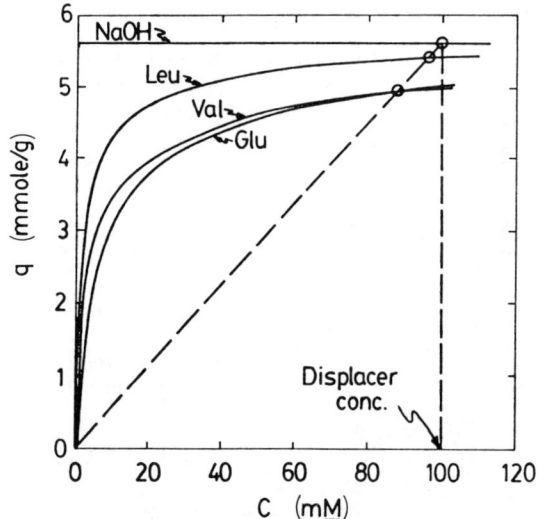

Figure 8. Pure component isotherms describing the uptake of amino acids by the hydrogen form of Dowex 50-X8 resin.

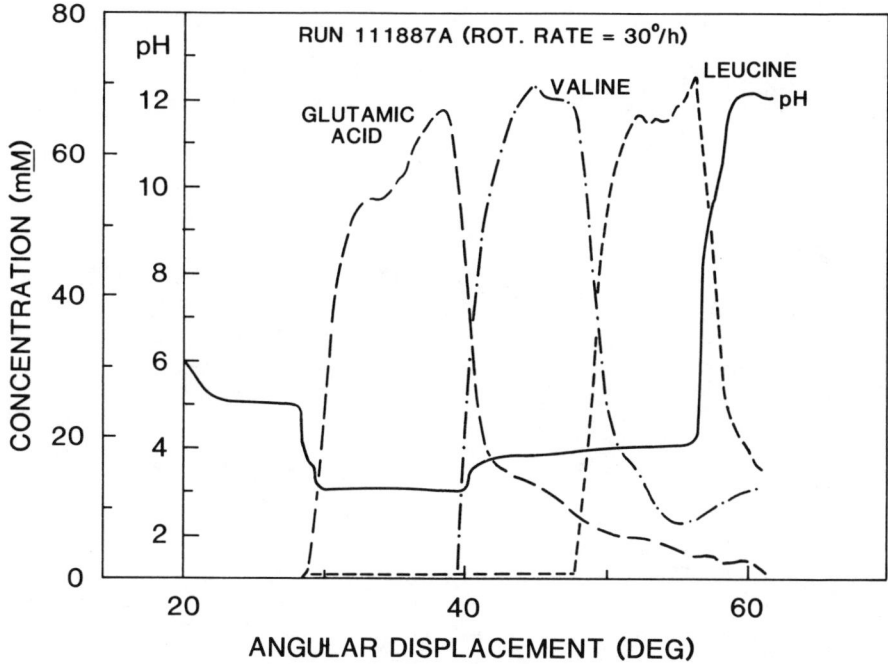

Figure 9. Displacement chromatography of an amino acid mixture by CAC using a 0.1 N NaOH displacer solution.

Chapter 11

Preparative Liquid Chromatography of Biomolecules — New Directions

Steven M. Cramer
Guhan Subramanian

The field of preparative chromatography has seen a
period of rapid growth in the last decade. The
development of novel stationary phases with
increased selectivities has enabled the separations
of complex mixtures of biomolecules. In addition,
the use of preparative chromatographic columns in
alternative modes of operation have greatly
increased the throughputs attainable with these
systems. In this review, we present some of the
recent advances in the use of liquid chromatography
for preparative bioseparations and discuss various
high throughput engineering approaches to column
operation such as step gradient and displacement
chromatography.

INTRODUCTION

The separation of biomolecules from the complex mixtures
obtained in biotechnology processes is one of the major engineering challenges facing the biotechnology industry today. The
requirements placed on the purity of biopharmaceutical products
have greatly increased the importance of liquid chromatography
in bioseparation processes(1).

High performance liquid chromatography (HPLC) has become
the premier analytical technique for biomolecules due to the
development of highly efficient and selective stationary phases
along with the advent of sophisticated instrumentation(2). In
response to the urgent need for high purity bioproducts, advances in analytical HPLC are now being exploited for increasingly
larger scale bioseparations(3,4). While the goal of analytical
chromatography is the analysis of the composition of a given
feed mixture, the objective in preparative chromatography is to

isolate and purify the bioproducts(1,5,6,7). This fundamental difference is reflected in the relative importance of various chromatograhic parameters. Whereas emphasis is placed on resolution and analysis time in analytical separations, the critical parameters in preparative chromatography are the amount of material isolated per unit time and the purity obtained(1,8). Since the objectives and constraints are different, the direct scale-up of analytical separations is generally not sufficient for process-scale chromatography. Thus, it is critical that novel engineering approaches be employed to fully exploit the high selectivity obtained with analytical chromatography for process-scale bioseparations.

There have been several reviews on the use of pre parative chromatography for the separation of biomolecules(7-18). In this report, we will present some of the important recent bioseparations attained with preparative chromatography. In addition, we will discuss alternative modes of column operation such as frontal, step-gradient and displacement chromatography. Finally, some suggestions are provided for the design of an appropriate chromatographic strategy for a given bioseparation problem by employing combinations of the above mentioned techniques.

Although the techniques and modes of operation presented in this report represent the state-of-the-art of preparative chromatography of biomolecules, we should mention at the outset that this report has been limited to applications reported in the literature. In fact, many of the industrially significant applications are not reported in the literature until several years after the techniques are developed.

THEORETICAL MODEL

The movement of a solute through a chromatographic column can be described by the simple mass balance expression(19):

$$\frac{\partial c_i}{\partial t} + u_0 \frac{\partial c_i}{\partial z} + \frac{1-\varepsilon}{\varepsilon} \frac{\partial \bar{q}_i}{\partial t} - D_i \frac{\partial^2 c_i}{\partial z^2} = 0 \qquad i=1,2,\ldots,N \tag{1}$$

$$\frac{\partial \bar{q}_i}{\partial t} = k_i \left(q_i^* - \bar{q}_i \right) \tag{2}$$

where c_i = concentration of species i in the mobile phase

$\overline{q}_i$ = average concentration of species i in the stationary

phase

u_0 = interstitial velocity

ε = void fraction of the bed

q_i^* = stationary phase concentration of species i in the

absence of mass transfer limitations

D_i = axial dispersion coefficient for species i

k_i = effective mass transfer coefficient of species i

It has been established that the equilibrium adsorption of most substances in chromatographic systems can be described by the langmuir adsorption isotherm(20-22). For a multicomponent system, the langmuir isotherm for component i is given by:

$$q_i^* = \frac{a_i c_i}{1 + \sum\limits_{j=1}^{N} b_j c_j} \tag{3}$$

where a_i and b_i are the langmuir parameters for species i and N is the number of components in the mixture(21-23). The system of equations described by Equations (1) and (2) are coupled through this multicomponent adsorption isotherm.

ELUTION CHROMATOGRAPHY

The most common mode of column operation is elution chromatography(24). Under these conditions, a feed mixture is injected into the column inlet as a finite volume pulse. The feed components then migrate through the column at different speeds which are a function of the mobile phase velocity and the distribution of the compounds between the mobile and stationary phases(7,25). Elution chromatography can be carried out with either a constant mobile phase composition (isocratic) or a continuous gradient in the mobile phase composition.

The critical parameters in preparative chromatography are the amount of material isolated per unit time and the purity obtained. In this paper, throughput is defined as the amount of

material purified per unit time at a specified purity. The
interplay of various operating parameters on the throughput and
purity obtained in preparative elution chromatography have been
examined by many authors(5,6,8,11,26-34) and will not be
addressed in this paper.

Preparative elution chromatography is generally carried out
under mass and/or volume overloaded conditions in order to
increase product throughput(5,7,26-28). In volume overloading,
the sample concentration is maintained in the linear region of
the isotherm and the volume is increased until the throughput is
optimized(5,7,28). A fundamental problem with this technique is
the under-utilization of the column and the corresponding low
throughputs. In mass overloading, the sample concentration
is increased beyond the linear adsorption region(5,7,28,35).
While complete analytical solutions of the equations describing
elution chromatography are obtainable for linear chromatography,
numerical solutions are generally required for non-linear
chromatography(36-39). Figure 1 shows typical elution profiles
resulting from volume and mass overloaded conditions. In both
cases, the peak shapes deviate from the gaussian profiles
obtained in analytical chromatography. Under conditions of
volume overloading, the peaks have a flat top and are sym-
metrical in shape(5,28). On the other hand, with mass
overloading, the band profile becomes asymmetric, with self-
sharpening front and tailing rear boundaries for langmuirian
adsorption systems(36,40,41). An excellent discussion of
overloading in preparative chromatography is given by Knox and
Pyper(28) in which they establish that mass overloading results
in significantly higher production rates than volume
overloading(42,43). In fact, a combination of volume and mass
overloading is commonly used to maximize the throughput in preparative elution chromatography(7).

The throughput in overloaded preparative chromatography can
be further increased by employing a two-step
procedure(1,9,41,44). In the first step, the column is substantially overloaded to effect the crude separation of a large
amount of closely eluting components from other contaminants
which have significantly different retention. In the second
step, fractions containing the desired material can then be
pooled and re-chromatographed to obtain the bioproduct in pure
form. Figure 2 demonstrates how this strategy can be employed
in the purification of glycopeptide antibiotics(45).

For the preparative separation of components with relatively small separation factors, high column efficiencies are
required. While this is accomplished in analytical chromatography with the use of small adsorbent particles, the use of
small particles for large-scale systems is often impractical
(7,9,11,12). Operational difficulties along with economic

constraints have limited the usage of small particles to analytical and semi-preparative systems to date(11,12). In fact, under highly overloaded conditions, there is minimal advantage to using small particle adsorbents in large-scale elution systems(9,46). In order to obtain sufficient efficiency with larger adsorbent particles, it is then necessary to use longer columns(8). However, since the economics of operating sufficiently long preparative columns is prohibitive, preparative columns are often employed in the recycle mode(1,8,18). By recycling the column effluent or a portion thereof, the efficiency of these systems can be significantly increased as shown in Figure 3.

STATIONARY PHASES

Since high efficiency in preparative chromatographic systems is generally impractical, it is critical to choose high selectivity adsorbents for a given preparative bioseparation. The support materials generally employed for preparative chromatography are either surface-modified porous silica or various organic gels. There are many stationary phase systems currently available for use in preparative and process-scale chromatograhy and a thorough survey of available commercial packings is given by Unger and Janzen(47).

The adsorptive systems commonly employed in preparative chromatography are: ion exchange (IEC); reversed phase (RPLC); normal phase; hydrophobic interaction (HIC); affinity; and metal chelate interaction (MCIC) chromatography. Size exclusion chromatography (SEC) is also widely employed in preparative chromatography. In this report, however, we will limit ourselves to adsorptive chromatographic systems in keeping with the spirit of this book.

APPLICATIONS

Appendix 1 presents some of the important applications of preparative chromatography of biomolecules reported in the literature in the recent past. Bibliographies which cover earlier work in preparative chromatography are available from both Waters(48) and Varex(49) corporations. While every attempt has been made to include many of the important contributions to the field of preparative chromatography of biomolecules, the list in Appendix 1 is by no means complete.

For each separation, we list the purified bioproduct, column dimensions, sample amount, particle size, mode of operation, and stationary phase adsorbent. All separations employing similar stationary phase materials have been grouped together. The bioseparations presented in this table employ a wide variety of operating parameters. Particle diameters ranging from 5 to

105 microns have been used in columns with inner diameters ranging from 0.46 to 20 cm. The bioseparations presented here represent separations ranging from small peptides to large proteins, on the scale of milligrams to hundreds of grams. For each separation, the mode of operation is presented. The symbols i, g, r, s-g, and d represent isocratic, continuous gradient, recycle, step-gradient, and displacement chromatography, respectively.

IEC has been extensively employed for the preparative separations of proteins(16,50,51). The reader is referred to the review by Regnier(51) for an overview of ion-exchange chromatography of proteins. Figure 4 shows the preparative separation of 100 mg of a mixture of 21 bacterial ribosomal proteins using a TSK SP-5-PW cation exchange column with a linear salt gradient(52). The five unresolved proteins were subsequently purified by preparative reversed phase chromatography. This separation demonstrates the inherent resolving power of preparative chromatographic systems.

RPLC has been successfully employed for the purification of a variety of biomolecules(45,53,54). Figure 5 demonstrates the process-scale separation of 270 grams of β-lactam antibiotics using an octadecyl silica column with a methanol step-gradient(55). A 60 x 20 cm I.D. column packed with 55-105 micron supports was used in this separation which clearly demonstrates the throughputs that can be achieved in preparative chromatographic systems. While reversed phase chromatography has been extensively used for the separation of small biomolecules, it has only been of limited utility for the preparative separation of proteins due to the denaturation accompanying the separation process(50,54).

HIC has exhibited a rapid growth in its use for the purification of proteins. This process utilizes a decreasing salt gradient to selectively elute the hydrophobic solute molecules(50,56,57). Although both HIC and RPLC use hydrophic interactions to effect the separation, HIC does not produce the degree of denaturation usually accompanying separations by RPLC(47,50,57). Goheen and Matson(58) used a decreasing linear gradient of ammonium sulfate in a 7.5 x 0.75 cm TSK-phenyl-5PW column to purify γ-globulin from human serum. Kato et al.(59) have used HIC for the purification of 200 mg of lipoxidase.

Normal phase chromatography has also been employed for the preparative separation of biomolecules(47,53). Kubo et al.(60) have separated diastereomeric ipsdienol derivatives using silica columns in the recycle mode with a mobile phase of n-pentane-acetone (100:0.2).

Affinity chromatography is a uniquely selective technique

for the purification of biomolecules(61-64). This method employs either biospecific ligands or synthetic dyes to selectively remove the bioproduct. The reader is referred to Janson(61) for a complete description of the process. Figure 6 demonstrates the affinity purification of vancomycin from clarified fermentation broth using an affigel D-ala-D-ala stationary phase(65). The antibiotic was subsequently eluted with an ammonia-acetontrile wash. Affinity chromatography is readily scaled-up to process-scale separations. Folena-Wasserman et al.(65) have purified 36 grams of the antibiotic aridicin from 4 liters of clarified fermentation broth using the above mentioned dipeptide affinity support. Affinity chromatography can also result in the concentration of the bioproduct, making this technique particularly attractive for the downstream processing of dilute fermentation mixtures.

Since its introduction by Porath et al.(66), metal chelate interaction chromatography (MCIC) has gained wide acceptance as a useful technique for the separation of a wide variety of biopolymers(67). Andersson and Porath(68) have recently used an iminodiacetate-ferric column in the purification of ovalbumin. Weerasinghe et al.(69) have employed zinc chelate chromatography to isolate coagulation factor XII from human plasma.

In addition to the above mentioned chromatographic stationary phases, there has been much activity in the development of alternative stationary and mobile phases for preparative separations. Kant and Hallen(70) have used a "homemade" magnesium containing stationary phase, magnammsil, for the separation of cerebrosides. Miguel et al.(71) have employed ion pairing agents in reversed phase chromatography for the preparative separation of aminoglycoside antibiotics. Chapple and Ellis(72) have separated caffeoylated phenylethanol glycosides using a polyamide-6 stationary phase with a linear ethanol gradient. Chiral stationary and mobile phases have been extensively employed for analytical separations of enantiomeric biomolecules(73-75) and are expected to be useful in preparative separations in the near future.

Although the list of applications presented in Appendix 1 is not complete, it serves to illustrate the wide range of biomolecules which have been separated by a variety of preparative chromatographic techniques.

ALTERNATIVE CHROMATOGRAPHIC MODES

Preparative elution chromatography is generally carried out under mass overloaded conditions which results in self-sharpening fronts and tailing rears for langmurian adsorption systems as shown in Figure 1(7,41). This pronounced tailing

leads to significant mixing of the feed components with concom-
mittant reduced production of the bioproduct. While recycle
chromatography can increase the production rate in elution
systems, it can be both expensive and cumbersome(7).

On the other hand, the throughputs in these chromatographic
systems can be readily increased by operating the columns in
alternative modes(18,76). In this section, we will focus on
various methods which involve operational changes in column
inlet conditions at specific times. Specifically, we will
discuss frontal, step-gradient, displacement and multimodal
chromatography.

A convenient way to describe these modes of column operation is to examine the initial and boundary conditions which,
along with the material balance equations, completely describe
the process. Appendix 2 presents the initial and boundary conditions for elution, frontal, step-gradient, and displacement
chromatography. In all cases, the column is initially
equilibrated with the carrier mobile phase. In frontal chromatography, the feed mixture is then continuously pumped into the
column(7,77). Under these conditions, the breakthrough of the
weakly adsorbing components occur rapidly, while the more
strongly adsorbing components are retained in the column for a
longer period of time. Thus, biological mixtures containing
species of widely differing affinities for a given stationary
phase can be separated by this relatively simple method. The
retained products can then be further purified by either step-
gradient and/or displacement chromatography as described below.
In point of fact, in all preparative chromatographic separations, the introduction of the feed mixture is always carried
out in the frontal chromatographic mode. When properly employed,
this mode can be exploited to separate the bulk of the nonad-
sorbing impurities prior to any additional changes in the inlet
boundary conditions.

Cramer and Horvath have successfully used frontal chromatography in concert with displacement chromatography for the
purification of peptides(78). Figure 7 shows the separation of
5 grams of a peptide mixture obtained from an immobilized car-
boxypeptidase Y catalyzed synthesis(78). In this separation,
the bulk of the less adsorbing L-methioninamide was removed
during the frontal chromatographic step, while the remaining
peptides were subsequently purified and concentrated by displacement chromatography on the same column.

The initial phase of step-gradient chromatography is identical to that of frontal chromatography, as shown in Appendix 2.
The feed is pumped into the column under conditions where the
compound of interest is strongly adsorbed. Following the removal of the less adsorbing species, the mobile phase composition

is abruptly changed to facilitate the elution of the compound of interest(8,18,76). In fact, several step-changes in mobile phase compositions can be sequentially employed to selectively remove one component at a time. Sakuma and Motomura have used multi-step gradients to purity saponins from B. falcatum as shown in Figure 8(79). Folena-Wasserman et al.(80) have employed this technique in the purification of malaria vaccine candidates from E. coli.

Step-gradient can also be used in concert with other chromatographic modes. Lee et al.,(81) have employed tandem separation schemes involving frontal chromatography followed by stepwise desorption or displacement for the purification of protein mixtures. Cramer and Horvath(78) have used step-changes in both organic solvent and column temperature to facilitate the purification of N-benzoyl-L-arginyl-L-methionyl-L-leucinamide by displacement chromatography as shown in Figure 9.

Displacement chromatography is a powerful preparative technique for the separation of biomolecules due to the high throughput and product purity associated with the process(7,76,82). In displacment chromatography, a front of displacer solution traveling behind the feed drives the separation of the feed components into adjacent pure zones which move at the same velocity as the displacer front. The separation is based on competition of the compounds for adsorption sites on the stationary phase and the process takes advantage of the nonlinearity of the isotherms(83,84). The boundary conditions describing this process are given in Appendix 2. The displacer is selected so as to have a higher affinity for the stationary phase than any of the feed components. The final concentration of each zone is determined solely by its adsorption isotherm and the concentration and isotherm of the displacer as shown in Figure 10. Since displacement chromatography occurs at concentrations corresponding to the non-linear region of the isotherms, a larger feed can be separated on a given column with the purified components recovered at significantly higher concentrations(82,83). Upon completion of the displacement process, the displacer is removed from the column by the passing of a suitable regenerant solution.

Although the technique of displacement chromatography has been known for 45 years, the potential of this technique for preparative bioseparations was not realized until the recent work by Horvath and co-workers(83-88) who have employed HPLC columns in the displacement mode for the preparative separation of a variety of biomolecules.

Cramer and Horvath(78) have employed displacement chromatography for the simultaneous concentration and purification of peptides as shown in Figures 7 and 9. We have recently employed

organic modifiers in the carrier along with elevated column temperatures to effect the efficient separation of relatively
hydrophobic peptides by displacement chromatography with
reversed phase HPLC as shown in Figure 11 (82). Furthermore,
under these conditions, elevated flow rates could be employed
with minimal effect on product purity resulting in the simultaneous purification and concentration of 40 mg of peptides in 8.3
minutes on an analytical reversed-phase column. While this does
not include the time required for regeneration and re-
equilibration with the carrier, it is expected that the total
cycle time for this separation would be less than 20 minutes.
The antibiotic cephalosporin C has been isolated from the
impurities in a fermentation broth by displacement chroma-
tography(82). We are presently conducting a detailed investigation of the scale-up of this bioseparation (89). Displacement
chromatography has also been successfully employed for the purification of proteins(82,90,91). Figure 12 shows the separation
of cytochrome c and lysozyme by displacement chromatography on a
cation exchanger using a cationic polymer as the displacer(82).
We have recently extended this separation to the simultaneous
purification and concentration of a mixture of five
proteins(92).

    The scale-up of the displacement process with respect to
stationary phase particle size, column dimensions, feed load,
volumetric flow rate, and molecular dimensions of the feed and
displacer components are expected to affect the "sharpness" of
the boundaries separating adjacent displacment zones and consequently the throughput and purity attained in these scaled-up
systems. In order to facilitate the scale-up of the process, we
have recently developed a mathematical model of displacement
chromatography which include the effects of axial dispersion and
finite mass transfer(19). Figure 13 demonstrates the effect of
finite mass transfer on the effluent displacement profile. For
a Stanton number of 1600, the displacement effluent profile
approaches that obtained under ideal chromatographic
conditions(19). When the Stanton number is decreased to 200,
the concentration shock waves separating the displacement zones
become more diffuse, resulting in increased zone overlap as
illustrated in the figure. We have also examined the influence
of mass transfer limitations on product throughput as a function
of particle diameter, interstitial velocity, and solute diffusivity as shown in Figure 14(19). The model predicts that the
throughput can be dramatically increased by operating at elevated velocities when small particle diameter adsorbents are
employed. In fact, this prediction has been confirmed by our
experimental studies on the displacement chromatography of
peptides(82). As the molecular dimensions of the solute
increase, however, the throughput becomes an increasingly
stronger function of the particle diameter. Thus, it is
expected that small particles will be necessary for the separation of macromolecules by displacement chromatography.

While liquid chromatography is widely employed for bioseparations, there is presently some confusion regarding
appropriate engineering strategies for optimizing the throughput
in these systems. When high selectivity can be achieved for the
bioproduct of interest, step-gradient or "on-off" chromatography
can be employed for these difficult high throughputs. On the
other hand, for complex mixtures containing closely eluting compounds, it has been established that displacement chromatography
can be successfully employed for these difficult preparative
separations(82,92). Although displacement chromatography is
presently a relatively unknown technique, we believe this mode
of chromatography will become the separation method of choice
for difficult bioseparation problems due to its ability to
separate large quantities of closely eluting bioproducts at high
levels of purity. In fact, the high throughput and purity
attained in displacement chromatography can often be further
enhanced by employing frontal chromatography, column washes
and/or step gradients in concert with the displacement chromatographic step(81).

CONCLUSIONS

In this brief review, we have attempted to present the
reader with some of the recent developments in the rapidly
growing field of preparative chromatography of biomolecules.
While preparative chromatography has been employed for separations ranging from milligrams to hundreds of grams to date, it
is expected that advances in the technology will dramatically
increase the throughput and purity attainable with these systems
in the near future. Since high efficiency in preparative chromatographic systems is often impractical, it is critical to choose
high selectivity adsorbents for a given preparative separation.
If a chromatographic system can be established for a given
bioseparation which results in large separation factors of the
feed components, then frontal and/or step gradient chromatography can be employed to produce high product
throughput(18,76). On the other hand, for chromatographic
separations involving relatively small separation factors,
displacement chromatography can be used to effect the simultaneous purification and concentration of the bioproduct. It is
expected that intense commercial competition along with a high
level of research activity will continue to spur rapid advances
in the field of preparative chromatography of biomolecules. In
addition, the use of sequential chromatographic steps in a
single column, will result in efficient use of the chromatographic bed, and will further improve the throughput and economics of preparative liquid chromatography.

REFERENCES

1. Colin, H.; Lowy, G.; Cazes, J. J. Am. Lab., 1985, 17(5),
    136.

2. Henschen, A.; Hupe, K.P.; Lottspeich, F.; Voelter, W. in
    High Performance Liquid Chromatography in Biochemistry;
    VCH, FRG, 1985.

3. Kitka, E.J. Sepn. and Purif. Methods, 1984, 13(2), 109.

4. Verzele, M.; Dawaele, C.; Van Dijck, J.; Van Haver, D. J.
    Chromatogr., 1982, 249, 231.

5. Scott, R.P.W.; Kucera, P. J. Chromatogr., 1976, 119, 467.

6. Hupe, K.P.; Lauer, H.H. J. Chromatogr., 1981, 203, 41.

7. Guiochon, G.; Katti, A. Chromatographia, 1987, 24, 165.

8. McDonald, P.D.; Bidlingmeyer, B.A. in Preparative Liquid
    Chromatography; Bidlingmeyer, B.A., Ed., Elsevier,
    Amsterdam, 1987, p. 1.

9. Sitrin, R.; dePhillips, P.; Dingerdissen, J.; Erhad R.;
    Filan, J. LC-GC, 1986, 4(6), 530.

10. Verzele, M.; Geeraert, E. J. Chromatogr. Sci., 1980, 18,
    559.

11. Heckendorf, A.H.; Ashare, E.; Raush, C. in Purification of
    Fermentation Products; LeRoith, D., Shiloach, J., and
    Leahy, T.J., Eds., ACS Symp. 271, 1985, 91.

12. Focus, Anal. Chem., 1985, 57(9), 998A.

13. Janson, J.C.; Hedman, P. Adv. Biochem. Eng., 1982, 25, 43.

14. Mazsaroff, I.; Regnier, F.E. J. Liq. Chrom., 1986, 9(12),
    2563.

15. Barker, P.E.; Ganetsos, G. Sepn. & Purif. Methods, 1988,
    17(1), 1.

16. Regnier, F.E.; Mazsaroff, I. Biotech. Progress, 1987,
    3(1), 22.

17. Verzele, M.; Dawaele, C. in Preparative HPLC, A Practical
    Guideline; TEC, Gent, Belgium, 1985.

18. Wankat, P.C. in <u>Large-Scale Adsorption and Chromatography</u>, Vol. 1 and 2, CRC Press, Boca Raton, FL, 1986.

19. Phillips, M.W.; Subramanian, G.; Cramer, S.M. "A theoretical optimization of operating parameters in non-ideal displacement chromatography," <u>J. Chromatogr.</u>, 1988, <u>454</u>, 1.

20. Condor, J.R.; Young, C.L. <u>Physicochemical Measurements by Gas Chromatography</u>, John Wiley & Sons, NY, 1979.

21. Jacobson, J.; Frenz, J.; Horvath, C. <u>J. Chromatogr.</u>, 1984, <u>316</u>, 53.

22. Ruthven, D.M., <u>Principles of Adsorption and Adsorption Processes</u>, John Wiley & Sons, NY, 1984.

23. Jacobson, J.; Frenz, J.; Horvath, C. <u>Ind. Eng. Chem. Res.</u>, 1987, <u>26</u>, 43.

24. Ludge, S.R.; Ladish, M.R. in <u>Separation, Recovery and Purification in Biotechnology</u>; Asenjo, J. A.; Hong, J., Eds., ACS Symp. Series, 314, 1986, 122.

25. Snyder, C.R.; Kirkland, J.J. <u>Introduction to Modern Liquid Chromatography</u>, John Wiley & Sons, NY, 1979.

26. Colin, H. <u>Sepn. Sci. & Tech.</u>, 1987, <u>22</u>(8-10), 1851.

27. Cretier, G.; Rocca, J.L. <u>Sepn. Sci. & Tech.</u>, 1987, <u>22</u>(8-10), 1881.

28. Knox, J.H.; Pyper, H.M. <u>J. Chromatogr.</u>, 1986, <u>363</u>, 1.

29. Hupe, K.P.; Hoffmann, B.; <u>Sepn. Sci. & Tech.</u>, 1987, <u>22</u>(8-10), 1869.

30. DeJong, A.W.; Smit, J.C.; Poppe, H.; Kraak, J.C. <u>Anal. Proc. (Lond.)</u>, 1980, <u>17</u>, 508.

31. Janson, J.-C.; Hedman, P. <u>Biotech. Progress</u>, 1987, <u>3</u>(1), 9.

32. Newburger, J.; Liebes, L.; Colin, H.; Guiochon, G. <u>Sepn. Sci. & Tech.</u>, 1987, <u>22</u>(8-10), 1933.

33. Meyer, V.R. <u>Sepn. Sci. & Tech.</u>, 1987, <u>22</u>(8-10), 1909.

34. Guiochon, G.; Colin, H. <u>Chrom. Forum</u>, 1986, <u>1</u>(3), 21.

35. Poppe, H.; Kraak, J.C. <u>J. Chromatogr.</u>, 1983, <u>255</u>, 395.

36. Rouchon, P.; Schonauer, M.; Valentin, P.; Guiochon, G.
     Sepn. Sci. & Tech., 1987, 22(8-10), 1793.

37. Said, A.S. Sepn. Sci. & Tech., 1981, 16(2), 113.

38. Gareil, P.; Rosset, R. Sepn. Sci. & Tech., 1987, 22(8-10),
     1953.

39. Jacob, L.; Guiochon, G. Chromatogr. Rev., 1971, 14, 77.

40. Jaulmes, A.; Vidal-Madjar, C.; Colin, H.; Guiochon, G. J.
     Phys. Chem., 1986, 90, 207.

41. Guiochon, G., New Directions in Chemical Analysis; Shapiro,
     B.L., Ed., Texas A&M Univ. Press, Texas, 1985, p. 84.

42. Gareil, P.; Durieux, C.; Rosset, R. Sepn. Sci. & Tech.,
     1983, 18, 441.

43. Cretier, G.; Rocca, J.L. Chromatographia, 1985, 20, 461.

44. Guiochon, G.; Roz, B; Bonmati, R; Hagenbach, P; Valentin,
     P. J. Chromatogr. Sci. 1976, 14, 367.

45. Sitrin, R.D.; Chan, G.; dePhillips, P.; Dingerdissen, J.;
     Valenta, J; Snader, K. Purification of Fermentation
     Products; LeRoith, D.; Shiloach, J.;   Leahy, T.J.,
     Eds., ACS Symp. 271, 1985, 71.

46. deJong, A. W.; Poppe, H.; Kraak, J. C. J. Chromatogr.,
     1981, 209, 432.

47. Unger, K.K.; Janzen, R. J. Chromatogr., 1986, 373, 227.

48. Preparative LC Bibliography, Publication T78, Waters
     Associates, Milford, MA, 1984.

49. Cazes, J., Preparative-Scale HPLC, A Bibliography,
     Publication V-100, Varex Corp., Rockville, MD, 1985.

50. Regnier, F.E. Science, 1983, 222, No. 4621, 245.

51. Regnier, F.E. Anal. Biochem., 1982, 126, 1.

52. Capel, M.; Datta, D.; Nierras, C.R.; Craven, G.R. Anal.
     Biochem., 1986, 158, 179.

53. Wehrli, A., in Preparative Liquid Chromatography,
     Bidlingmeyer, B.A., Ed., Elsevier, Amsterdam, 1987, p. 153.

54. Hancock, W.S.; Prestidge, R.L., in Preparative Liquid
     Chromatography, Bidlingmeyer, B.A., Ed., Elsevier,
     Amsterdam, 1987, p. 203.

55. Cantwell, A.M.; Calderone, R.; Sienko, M. J. Chromatogr., 1984, 316, 133.

56. Hofstee, B.H.J. J. Macromol. Sci., 1976, A10, 111.

57. Potter, R.L.: Lewis, R.V. High Performance Liquid Chromatography-Advances and Perspectives, Volume IV, Academic Press, Orlando, FL, 1986, p. 1.

58. Goheen, S.C.; Matson, R.S. J. Chromatogr., 1985, 326, 235.

59. Kato, Y.; Kitamura, T.; Hashimoto, T. J. Chromatogr., 1985, 333, 202.

60. Kubo, I.; Komatsu, S.; Iwagawa, T.; Wood, D.L. J. Chromatogr., 1986, 363, 309.

61. Janson, J.C. Trends in Biotech., 1984, 2(2), 31.

62. Carr, P.W.; Bergold, A.F.; Hanggi, D.A.; Muller, A.J. Chromatogr. Forum, 1986, 1 (3), 31.

63. Clonis, Y.D.; Jones, K.; Lowe, C.R. J. Chromatogr., 1986, 363, 31.

64. Dean, P.D.G.; Watson, D.N. J. Chromatogr., 1979, 165, 301.

65. Folena-Wasserman, G.; Sitrin, R.D.; Chapin, F.; Snader, K.M. J. Chromatogr., 1987, 392, 225.

66. Porath, J.; Carlsson, J.; Olsson, I.; Belfrage, G. Nature, 1975, 258, 598.

67. Sulkowski, E., in Protein Purification: Micro to Macro, Burgess, R., Ed., A.R. Liss, N.Y., 1987, p. 149.

68. Andersson, L.; Porath, J. Anal. Biochem., 1986, 154, 250.

69. Weerasinghe, K.M.; Scully, M.F.; Kakkar, V.V. Biochimica et. Biophysica Acta., 1985, 839, 57.

70. Kant, K.; Hallen, R.K. Anal. Biochem., 1985, 147, 455.

71. deMiguel, I.; Puech-Costes, E.; Samain, D. J. Chromatogr., 1987, 407, 109.

72. Chapple, C.S.; Ellis, B.E. J. Chromatogr., 1984, 285, 171.

73. Ravichandran, K.; Rogers, L.B. J. Chromatogr., 1987, 402, 49.

74. Abidi, S.L. J. Chromatogr., 1987, 404, 133.

75. Pirkle, W.H. in Chromatography and Separation Chemistry;
    Ahuja, S., Ed., ACS Symp. 297, 1986, p. 101.

76. Horvath, C.; Lee, A.; Liao, A.; Velayudhan, A.; Eng.
    Foundation IV Conference on Recovery of Bioproducts,
    Hawaii, April 17-22, 1988.

77. Karger, B.L., Snyder, L.R., Horvath, C. An Introduction to
    Separation Science, John Wiley & Sons, NY, 1973.

78. Cramer, S.M.; Horvath, C. Prep. Chromatogr., 1988, 1, 29.

79. Sakuma, S.; Motomura, H. J. Chromatogr., 1987, 400, 293.

80. Folena-Wasserman, G.; Inacker, R.; Cohen-Silverman, C.;
    Rosenberg, M. in Protein Purification: Micro to Macro;
    Burgess, R., Ed., A.R. Liss, Inc., NY, 1988, p. 337.

81. Lee, A. L.; Liao, A. W.; Horvath, C. J. Chromatogr., 1988,
    443, 31.

82. Subramanian, G.; Phillips, M.W.; Cramer, S.M. J.
    Chromatogr., 1988, 439, 341.

83. Horvath, C.S. in The Science of Chromatography, Bruner,
    F., Ed., Elsevier, Amsterdam, 1985, p. 179.

84. Frenz, J.; Horvath, C. AIChE J., 1985, 31, 400.

85. Kalasz, H.; Horvath, C. J. Chromatogr., 1982, 239, 423.

86. Horvath, C.S.; Nahum, A.; Frenz, J.H. J. Chromatogr.,
    1981, 218, 365.

87. Kalasz, H.; Horvath, C. S. J. Chromatogr., 1981, 215, 295.

88. Cramer, S.M.; El Rassi, Z.; Horvath, C. J. Chromatogr.,
    1987, 394, 305.

89. Phillips, M.W.; Cramer, S.M.; "An Investigation of Scale-Up
    in Displacement Chromatography", in preparation.

90. Liao, A.W.; El Rassi, Z.; LeMaster, D.M.; Horvath, C.S.
    Chromatographia, 1987, 24, 881.

91. Vigh, G.; Varga-Puchony, Z.; Szepesi, G.; Gazdag, M.; J.
    Chromatogr., 1987, 386, 353.

92. Subramanian, G.; Cramer, S.M.; "Displacement Chromatography
    of Protein Mixtures", in preparation.

93. Kling, G.J.; Perkins, L.M.; Cappiello, P.E.; Eisenberg, B.A. J. Chromatogr., 1987, <u>407</u>, 377.

94. Linde, S.; Welinder, B. S.; Hansen, B.; Sonne, O. J. Chromatogr., 1986, <u>369</u>, 327.

95. Charpentier, B.; Dingas, A.; Duval, D.; Emiliozzi, R. J. Chromatogr., 1986, <u>355</u>, 427.

96. Milhaud, J.; Gareil, P.; Rosset, R. J. Chromatogr., 1986, <u>358</u>, 284.

97. Shih, C.K.; Snavely, C.M.; Molnar, T.E.; Meyer, J.L.; Caldwell, W.B.; Paul, E.L. Chem. Eng. Prog., 1983, <u>79</u>, No. 10, 53.

98. Burgoyne, R.F., Bowles, D.K.; Heckendorf, A. Am. Biotech. Labs., March 1984.

99. Folena-Wasserman, G.; Poehland, B.L.; Yeung, E.W.K.; Staiger, D.; Killmer, L.B.; Snader, K.; Dingerdissen, J.J.; Jeffs, P.W. J. Antibiotics, 1986, <u>39</u>, No. 10, 1395.

100. Lowe, P.A.; Rhind, S.K.; Sugrue, R.; Marston, F.A.O. in Protein Purification: Micro to Macro; Burgess, R., Ed., A. R. Liss, Inc., NY, 1988, p. 429.

101. Bishop, C.A.; Harding, D.R.K.; Meyer, L.J.; Hancock, W.S.; Hearn, M.T.W. J. Chromatogr., 1980, <u>192</u>, 222.

102. Hearn, M.T.W.; Hancock, W. S., in Biological/Biomedical Applications of Liquid Chromatography II; Hawke, G. L., Ed.; Marcel Dekker, NY 1979.

103. Horvath, C.; Frenz, J.; El Rassi, Z., J. Chromatogr., 1983, <u>255</u>, 273.

104. El Rassi, Z., and Horvath, C., J. Chromatogr., 1983, <u>266</u>, 319.

105. Eble, J. E.; Grob, R. L.; Andle, P. E.; Cox, G. B.; Snyder, L. R., J. Chromatogr., 1987, <u>405</u>, 31.

106. Gareil, P.; Salinier, G.; Caude, M; Rosset R., J. Chromatogr., 1981, <u>208</u>, 365.

107. Nick, H. P.; Wettenhall, R. E.; Hearn, M.T.W.; Morgan, F. J., Anal. Biochem., 1985, <u>148</u>, 93.

108. Hui, K.S.; Hui, M.; Chiu, F.C.; Banay-Schwartz, M.; Deguzman, T.; Sacchi, R.S.; Lajtha, A. Anal. Biochem., 1986, <u>153</u>, 230.

109. Folena-Wasserman, G.; Inacker, R.; Rosenbloom, J. J. Chromatogr., 1987, 411, 345.

110. Becker, C. R.; Efcavitch, J. W.; Heiner; C. R.; Kaiser, N. F., J. Chromatogr. 1985, 326, 293.

111. Koeppe, R. E.; Weiss, C. B., J. Chromatogr., 1981, 208, 414.

112. Stampe, D.; Wieland, B.; Kohle, A. J. Chromatogr., 1986, 363, 101.

113. Nakamura, E.; Kato, Y., J. Chromatogr., 1985, 333, 29.

114. Schmuck, M. L.; Gooding, D. L.; Gooding, K. M., J. Chromatogr., 1986, 359, 323.

115. Muto, N.; Tan, L.; J. Chromatogr., 1985, 326, 137.

116. Strickler, M.P.; Gemski, M.J. J. Liq. Chromatogr., 1986, 9, 1655.

117. Deschamps, J. R.; Hildreth, J.E.K.; Deer, D.; August, J. T., Anal. Biochem., 1985, 147, 451.

118. Leaver, G.; Conder, J. R.; Howell, J. A., Sepn. Sci. & Tech., 1987, 22(8-10), 2037.

119. Scott, H.; Sepn. Sci. & Tech., 1987, 22(8-10), 2061.

120. Schmerr, M.J.F.; Goodwin, K. R.; Lehmkuhl, M. D.; Cutlip, R. C., J. Chromatogr. 1985, 326, 225.

121. Radulovic, L. L.; Kulkarni, A. P., Biochem. and Biophy. Res. Comm., 1985, 128, 75.

122. Casagli, M. C.; Borri, M. G.; D'Ettore, C.; Baldari, C.; Galeotti, C.; Bossu, P.; Ghiara, P.; Antoni, G., in Protein Purification: Micro to Macro; Burgess, R., Ed.; A. R. Liss, Inc., NY 1988, p. 421.

123. Torres, A. R.; Edberg, S. C.; Peterson, E. A., J. Chromatogr. 1987, 389, 177.

124. Torres, A. R.; Dunn, B. E.; Edberg, S. C.; Peterson, E. A., J. Chromatogr., 1984, 316, 125.

125. Torres, A. R.; Krueger, G. G.; Peterson, E. A., Anal. Biochem., 1985, 144, 469.

126. Hellerstein, M. K.; Sasak, V.; Ordovas, J., Munro, H. N., Anal Biochem. 1985, 146, 366

127. Kaminski, M.; Sledzinska, B.; Klauster, J., J. Chromatogr., 1986, 367, 45.

128. Risk, M.; Lin, Y. Y.; Ramanujam, V.M.S.; Smith, L. L.; Ray, S. M; Trieff, N. M., J. Chromatogr. Sci. 1979, 17, 400.

129. Sledzinska, B., J. Chromatogr. 1986, 303, 179.

130. Gooding, D. L.; Schmuck, M. L.; Nowlan, M. P.; Gooding, K. M., J. Chromatogr., 1986, 359, 331.

131. Fisher, A. M.; Yu, X. J.,; Bretaudiere, J. T.; Muller, D.; Bros, A.; Jozefonvicz, J. J Chromatogr., 1986, 376, 429.

132. Clonis, Y. D.; Jones, K.; Lowe, C. R.   J. Chromatogr., 1986, 363, 31.

133. Small, D.A.P.; Atkinson, T.; Lowe, C. R.   J. Chromatogr., 1983, 266, 151.

134. Laurent, P.; Miribel, L; Beinvenu, J.; Vallve, C.; Arnaud, P., FEBS. Letters, 1984, 168(1), 79.

135. Porath, J.; Belew, M., in Affinity Chromatography and Biological Recognition, I. M. Chaiken et al., Eds., Academic Press, Orlando, FL, 1983, 135.

136. Kikuchi, H.; Watanabe, M., Anal. Biochem., 1981, 115, 109.

137. Chadha, K. C.; P. M. Grob; Mikulski, A. J.; Davis, L. R.; Sulkowski, E., J. of Gen. Virology, 1979, 43, 701.

138. Onkubo, I.; Kondo, T.; Taniguchi, N., Biochemica et Biophysica Acta, 1980, 616, 89.

139. Hansson, H.; Kagedal, L., J. Chromatogr. 1981, 215, 333.

140. Sulkowski, E.; Vastola, K.; Oleszek, D.; Von Muenchhausen, W., Affinity Chrom. and related techniques, Gribnau, T.C.J.,; Visser, J.; Nivard, R.J.F., Eds., Elsevier, Amsterdam, 1982, p. 313.

141. Woodward, R. B.; in Ultra Resolution in Liquid Chromatography, Publ. AN72-123, Waters Associates, Milford, MA, 1972, p. 3.

142. Jobin Yvon, Division d'Instruments, S.A.

APPENDIX - 1

APPLICATIONS

PRODUCT	COLUMN DIMENSION (cm)	SAMPLE AMOUNT	PARTICLE SIZE (micron)	MODE OF OPERATION	STATIONARY PHASE	REFERENCE
β.lactam antibiotic	60x20	270g	55-105	i,s-g	C-18	55
amino-gylcoside antibiotics	25x0.46	1mg	10	i	C-18	71
indole-3-acetic acid	10x1	na	10	g	C-18	93
insulin	25x2.5	30mg	7	i	C-18	94
steroids	30x5.7	2g	35-75	i	C-18	95
filipin	50x0.9	4mg	10	i	C-18	96
saiko-saponins	100x11	10g	20	s-g	C-18	79
anti-biotics	50x1	200mg	37-53	i	C-18	97
anti-biotics	60x15	60g	37-53	i	C-18	97
polyoxin	50x2.5	100mg	25-40	s-g	C-18	45
glyco-peptides	50x2.5	2g	25-40	s-g	C-18	45
polymyxin	30x5.7	1g	55-105	g	C-18	98
poly-myxin	10x0.8	5mg	55-105	g	C-18	98
glyco-peptide antibiotics	30x2.2	140mg	na	s-g	C-18	99
calci-tonin	25x1	176mg	na	g	C-18	100

PRODUCT	COLUMN DIMENSION (cm)	SAMPLE AMOUNT	PARTICLE SIZE (micron)	MODE OF OPERATION	STATIONARY PHASE	REFERENCE
1-leu-(gly)$_3$	30x5.7	1g	75	i,r	C-18	101
insulin	na	2mg	na	i	C-18	102
dipep-tides	25x0.46	40mg	5	d	C-18	103
AMP	40x0.46	60mg	5	d	C-18	103
polym-myxin-B	25x0.46	150mg	5	d	C-18	86
guany-luridine	0.46	100mg	5	d	C-18	104
peptide mix	25x0.46	24mg	5	d	C-18	81
cephalo-sporin c	30x0.46	11ml	5	d	C-18	89
dipep-tides	25x2.2	5g	10	d	C-18	78
tripep-tides	40x0.46	22ml	5	d	C-18	78
tetrapep-tides	15x0.46	10ml	5	d	C-18	78
xanthines	15x0.46	27.5mg	5	i	C-8	105
polyenic antibiotic	47cmx1.25	17mg	5-20	i	C-8	106
protein S6	25x0.46	1.2mg	na	g	C-8	107
insulins	25x0.46	260mg	5	g	C-8	90
neuro-filament protein	25x1.0	5mg	na	g	C-4	108

PRODUCT	COLUMN DIMENSION (cm)	SAMPLE AMOUNT	PARTICLE SIZE (micron)	MODE OF OPERATION	STATIONARY PHASE	REFERENCE
circum- sporozoite protein	25x1	20mg	5	g	C-4	109
oligo- nucleotides	25x1	na	5	g	C-4	110
caffeoyl esters	21x2.5	1.0g	100-300	g	polyamide	72
cere-	40x2.5	10g	na	s-g	home made (magnammsil)	70
grami- cidins	300x0.78	100mg	37-75	i	phenyl silica	111
factor IX	50x18	na	na	i	IE	112
lipo- xidase	20x5.5	1g	20	g	IE	113
oval- bumin	25x2.1	3g	30-50	g	IE	114
cyto- chrome P-450	5x0.5	27.7mg	na	g	IE	115
mono- clonal antibodies	7.5x0.75	10mg	40	g	IE	116
IgG	7.5x0.75	19mg	na	g	IE	117
human malaria vaccine	20x2.5	150mg	na	s-g	IE	80
albumin	34x7.5	14g	150-300	s-g	IE	118
oligo- nucleotides	50x5	20g	na	s-g	IE	119
IgG	15x2.2	2 ml	na	s-g	IE	120

PRODUCT	COLUMN DIMENSION (cm)	SAMPLE AMOUNT	PARTICLE SIZE (micron)	MODE OF OPERATION	STATIONARY PHASE	REFERENCE
gluta-thione S-transferase	na	na	1mg	g	IE	121
inter-leukin	6.5x5	73.0mg	na	i	IE	122
inter-leukin	16x2.5	12.0mg	na	s-g	IE	122
β lacto-globulins	7.5x0.75	78mg	5	d .	IE	90
protein mix	25x0.46	12mg	5	d	IE	82
protein mix	25x0.46	24mg	5	d	IE	92
ribo-somal proteins	15x2.2	100mg	na	g	IE	52
protein mixture	7.5x0.75	25mg	na	d	IE	123
β-lacto-globulins	4.5x0.21	12.8mg	na	d	IE	124
gc-2-globulin	28x0.55	6ml	na	d	IE	125
glyco-protein	60x0.75	2.5mg	na	i	TSK G-3000 SW	126
lanato-sides	55x10	26g/gSiO$_2$	40-63	i	silica	127
ipsdie-nol	50x1	3g	5	i,r	silica	60
toxins	na	190mg	na	i	silica	128
glyco-sides	50x10	19.3g	40-63	i	silica	129

PRODUCT	COLUMN DIMENSION (cm)	SAMPLE AMOUNT	PARTICLE SIZE (micron)	MODE OF OPERATION	STATIONARY PHASE	REFERENCE
prosta-glandins	100x20	150g	na	i	silica	142
corti-costeroids	50x0.46	180mg	5	d	silica	85
human serum	7.5x0.75	2.0mg	na	g	HIC	58
lipox-idase	15x2.2	200mg	na	g	HIC	59
serum components	25x1	1 ml	30	g	HIC	130
thrombin	20x1	na	na	g	affinity	131
1-lactate dehydrogenase	200x15	1.8g	20	s-g	affinity	132
vanco-mycin	10x1	200mg	na	s-g	affinity	65
lactate dehydrogenase	30x2.5	13.9mg	40-63	s-g	affinity	133
glyco-protein	100x2.5	na	na	s-g	affinity	134
factor XII	na	320ml	na	i	metal chelate	69
serum components	26x1	4ml	na	s-g	metal chelate	135
non histonic proteins	7x1.3	3.2mg	na	s-g	metal chelate	136
inter-feron	5x0.9	12mg	na	s-g	metal chelate	137
nucleo-side diphosphatase	10x2	55mg	na	s-g	metal chelate	138

PRODUCT	COLUMN DIMENSION (cm)	SAMPLE AMOUNT	PARTICLE SIZE (micron)	MODE OF OPERATION	STATIONARY PHASE	REFERENCE
albumin	na	1.1g	na	s-g,g	metal chelate	139
albumin	na	3.6	na	s-g	metal chelate	139
oval-bumin	8x1	10mg	na	s-g	metal chelate	68
inter-ferons	8x0.9	5ml	na	s-g,g	metal chelate	140

APPENDIX - 2

INITIAL AND BOUNDARY CONDITIONS

In all cases, the column is initially equilibrated with the mixture The feed of N components at concentrations c_{1F}, $c_{2F}...c_{NF}$ is then introduced in the column for a time t_{feed} (except for frontal chromatography where the feed is continuously pumped through the column).

MODE	INITIAL & BOUNDARY CONDITIONS
ELUTION	$c_i = \overline{q}_i = 0$ @ $t=0$, $0 \leqslant z \leqslant L$ $i=0,1,2,...,N$
	$u_0 c_i = u_0 c_{iF} + D_i \left.\dfrac{\partial c_i}{\partial z}\right\|_{z=0}$ @ $z=0$, $0 \leqslant t \leqslant t_{feed}$ $i=1,2,...,N$
	$u_0 c_i = D_i \left.\dfrac{\partial c_i}{\partial z}\right\|_{z=0}$ @ $z=0$, $t > t_{feed}$ $i=1,2,...,N$
	$\left.\dfrac{\partial c_i}{\partial z}\right\|_{z=L} = 0$ @ $z=L$, $t \geqslant 0$ $i=1,2,...,N$
FRONTAL	$c_i = \overline{q}_i = 0$ @ $t=0$, $0 \leqslant z \leqslant L$ $i=1,2,...,N$
	$u_0 c_i = u_0 c_{iF} + D_i \left.\dfrac{\partial c_i}{\partial z}\right\|_{z=0}$ @ $z=0$, $t > 0$ $i=1,2,...,N$

MODE	INITIAL & BOUNDARY CONDITIONS
	$\left.\dfrac{\partial c_i}{\partial z}\right|_{z=0} = 0$ @ $z=L$, $t > 0$ $\quad i=1,2,\dots,N$
STEP GRADIENT	$c_i = \overline{q}_i = 0$ @ $t=0$, $0 < z < L$, $\quad i=1,2,\dots,N$
	$u_0 c_i = u_0 c_{iF} + D_i \left.\dfrac{\partial c_i}{\partial z}\right|_{z=0}$ @ $z=0$, $0 < t < t_{feed}$ $\quad i=1,2,\dots,N$
	$u_0 c_i = D_i \left.\dfrac{\partial c_i}{\partial z}\right|_{z=0}$ @ $t=0$, $t > t_{feed}$ $\quad i=1,2,\dots,N$
	$\left.\dfrac{\partial c_i}{\partial z}\right|_{z=L} = 0$ @ $z=L$, $t > 0$ $\quad i=1,2,\dots,N$

The boundary conditions are also accompanied by a step-change in the mobile phase conditions after time t_{feed}. This corresponds to a change in the adsorption condition. As a first order approximation, this disturbance in the adsorption conditions propagates through the column at some velocity, u_{eff}, which is a function of the dynamics of the system. The new mobile phase conditions will be felt at some position, x^*, in the column, corresponding to a time $t = \dfrac{x^*}{u_{eff}}$. Thus, we have a moving boundary condition in the column which can be represented as:

$$q_i^* = f_1(c_1, c_2, \dots c_N) \qquad @ \ t = t^*, \ x^* < x < L$$

$$q_i^* = f_2(c_1, c_2, \dots c_N) \qquad @ \ t = t^*, \ x < x^*$$

$$i = 1, 2, \dots, N.$$

MODE	INITIAL & BOUNDARY CONDITIONS

where f_1 and f_2 describe the multicomponent adsorption isotherms at the column inlet before and after the step gradient.

DISPLACEMENT　$\quad c_i = \overline{q}_i = 0$　　　　　　　@ t=0, 0 ⩽ z ⩽ L,

　　　　　　　　　　　　　　　　　　　　　　　　　i=1,2,...N

Following the introduction of the feed, a solution containing displacer at a concentration $c_{(N+1)F}$ is pumped into the column. The introduction of the displacer continues until time t_{displ}, the breakthrough time of the displacer at the column outlet.

$$u_0 c_i = u_0 c_i + D_i \left. \frac{\partial c_i}{\partial z} \right|_{z=0} \qquad @ \ z=0, \ 0 \ ⩽ \ t \ ⩽ \ t_{feed}$$

$$i=1,2,\ldots,N$$

and

$$c_{N+1} = 0 \qquad\qquad\qquad @ \ z=0, \ 0 \ ⩽ \ t \ ⩽ \ t_{feed}$$

$$u_0 c_i = D_i \left. \frac{\partial c_i}{\partial z} \right|_{z=0} \qquad @ z=0, \ t_{feed} \ < \ t \ < \ t_{displ}$$

$$i=1,2,\ldots,N$$

and

$$u_0 c_{N+1} = u_0 c_{(N+1)F} + D_{N+1} \left. \frac{\partial c_{N+1}}{\partial z} \right|_{z=0} @ \ z=0, \ t_{feed}$$

$$< \ t \ ⩽ \ t_{disp}$$

$$\left. \frac{\partial c_i}{\partial z} \right|_{z=L} = 0 \qquad\qquad @ \ z=L, \ t \ ⩾ \ 0,$$

$$i,1,2,\ldots,N$$

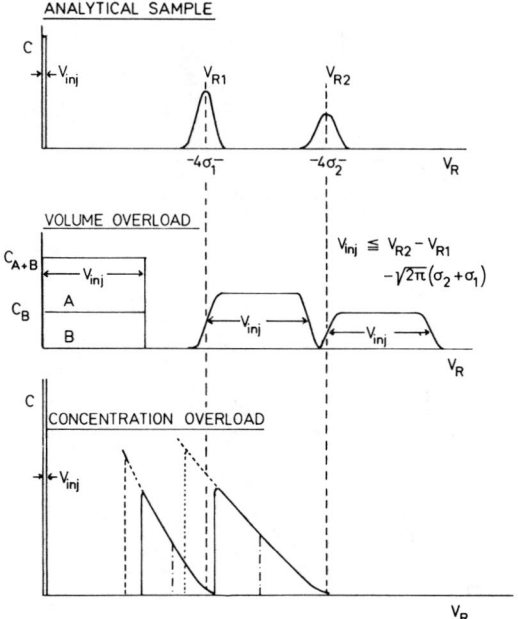

Figure 1. Effluent concentration profiles under conditions of volume and mass overload (reproduced with permission from reference #28, copyright 1986, Elsevier Science Publishers B.V.).

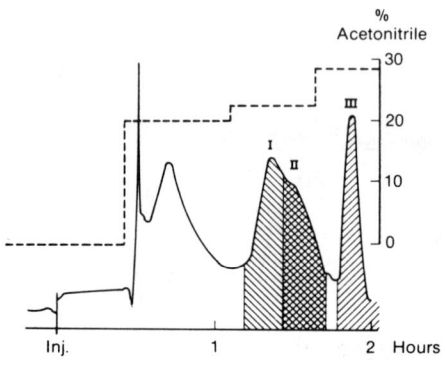

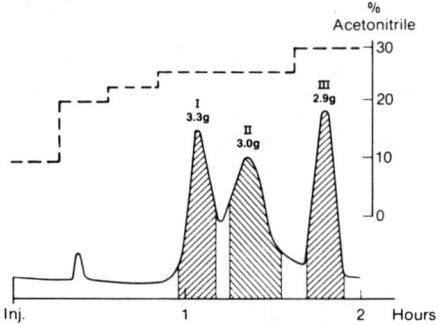

Figure 2.   Two-step procedure for separation of glycopeptide complex.   Column: Whatman Partisil Prep 40 ODS-3 (37-60 μm) packed in a 50x4.8cm column; flow rate: 250 ml/min, detection: 210 nm. a) separation of 50g of crude antibiotic complex.   Mobile phase: 0-28% linear gradient of ace tonitrile in 0.1M KH2PO4, pH 6.0.   b) rechromatography of pooled fractions obtained from the first step.   Mobile phase: 10-28% linear gradient of acetonitrile in 0.1M KH2PO4, pH 6.0 (reproduced with permission from reference #45, copyright 1985, American Chemical Society).

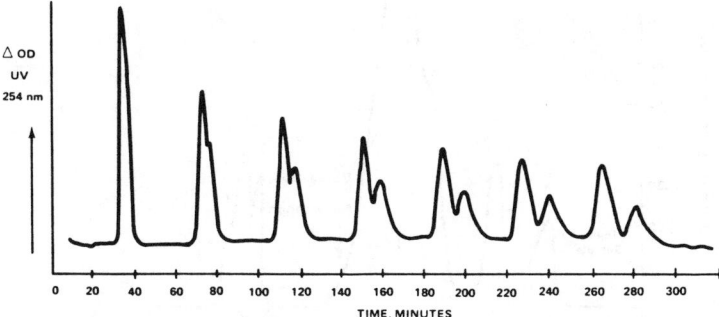

Figure 3. Recycle of normal- and neo-dicyanoheptamethyl-cobyrinates. Column: Five-24"x0.125" columns in series packed with Corasil 11 silica (37-75μm); mobile phase: hexane/methyl acetate/isopro-ponal/methanol-HCN (220:60:10:10); flow rate: 1 ml/min (adapted with permission from reference #141, copyright 1972, Waters Chromatography Division of Millipore).

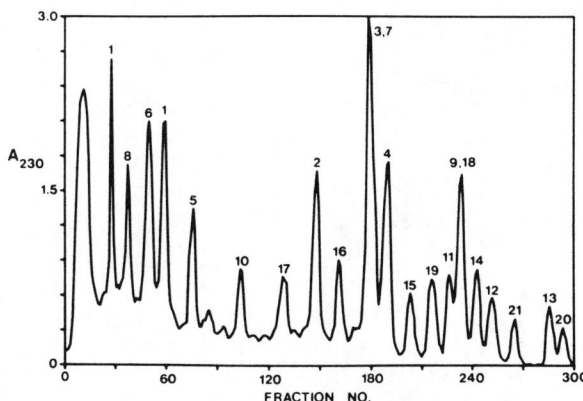

Figure 4. Preparative ion-exchange HPLC fractionation of 30S ribosomal proteins. Column: 150x21.5 mm TSK SP-5-PW; mobile phase: eluant A, 0.1 M KH$_2$PO$_4$, 8M urea, 0.5 mM dithiothreitol, pH 5.6; eluant B, same as eluant A with the addition of 1.0M KCl and readjustment of pH to 5.6 with H$_3$PO$_4$; gradient was a linear ramp from 100% eluant A to 35% eluant B in 600 min. (Reproduced with permission from reference #52, copyright 1986, Academic Press, Inc.).

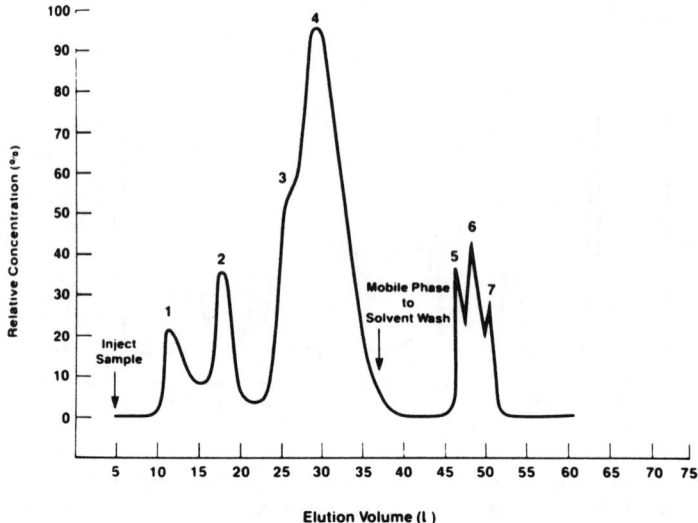

Figure 5.   Preparative separation of β-lactam antibiotic. Sample: 270g of crude antibiotic; column: 60x20cm column packed with C-18 silica (55-105 μm); mobile phase: water (reproduced with permission from reference #55, copyright 1984, Elsevier Science Publishers B.V.).

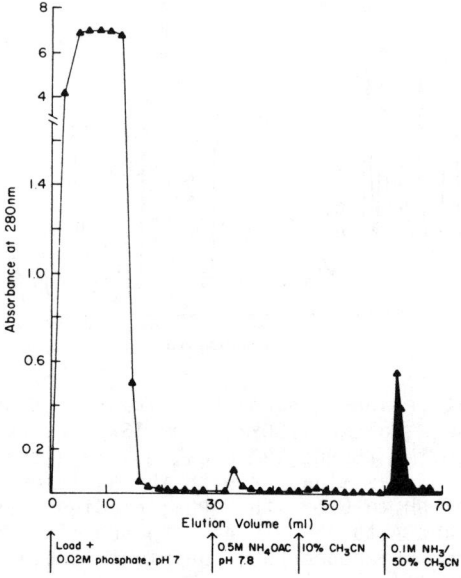

Figure 6.   Affinity purification of vancomycin from clarified fermentation broth.   Column: 10x1cm column packed with Affigel 10-D-Ala-D-Ala.   Elution of vancomycin is indicated by the shaded region. (reproduced with permission from reference #65, copyright 1987, Elsevier Science Publishers, B.V.)

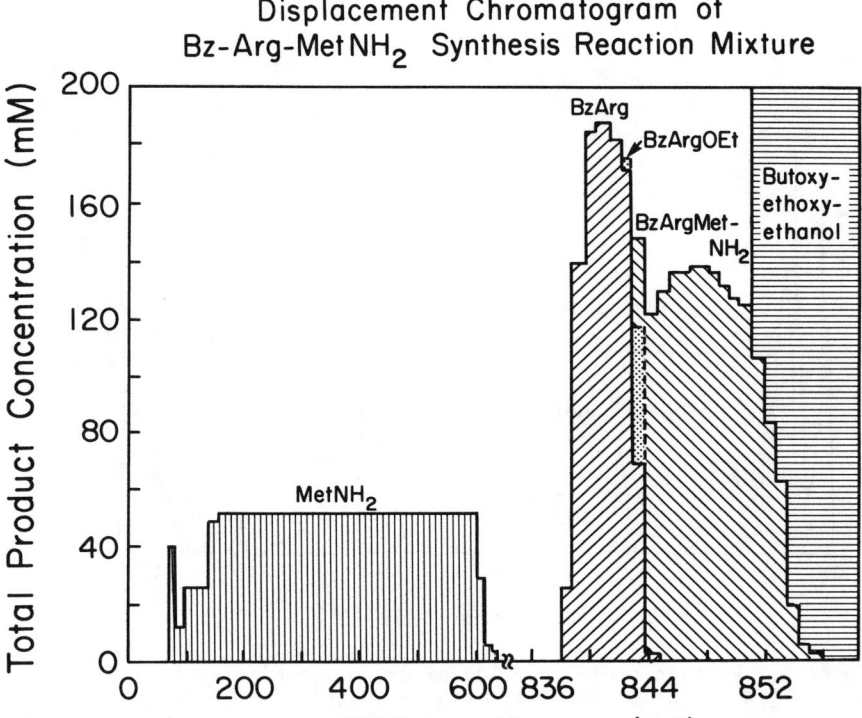

Figure 7. Displacement chromatography obtained in the purification of Bz-Arg-Met-NH2 from 500 ml of feed. Column: 250x22mm
in series with 400x4.6mm column packed with C-18 silica; carrier:
0.1M phosphoric acid, pH2.2; displacer: 247mM
2-(2-butoxyethoxy)ethanol in carrier; temperature: 22°C.
(Reproduced with permission from reference #78, copyright, 1988,
Gordon and Breach Science Publishers, Inc.).

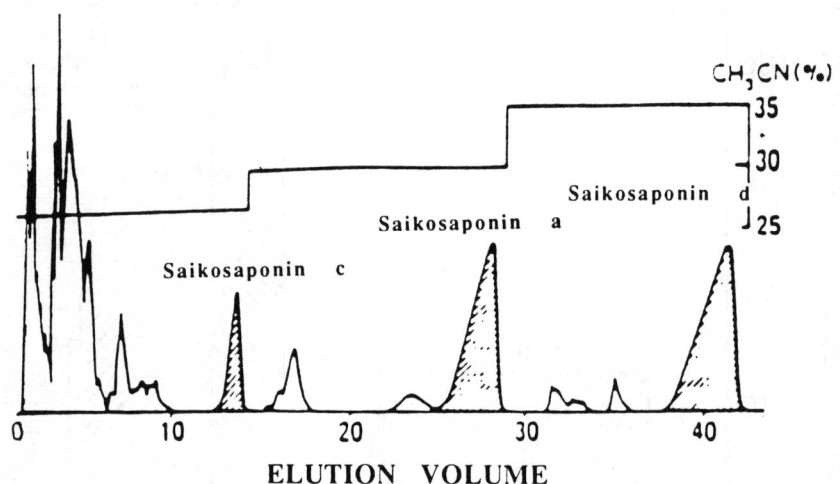

Figure 8. Elution profile of 10g of crude saponin fraction. Column: 918 x 110 mm ODS-Silica (20 μm); mobile phase: multiple step gradients of acetonitrile-water; flow rate: 210 ml/min. The elution volume is given in bed volumes. (Reproduced with permission from reference #79, copyright, 1987, Elsevier Science Publishers B.V.).

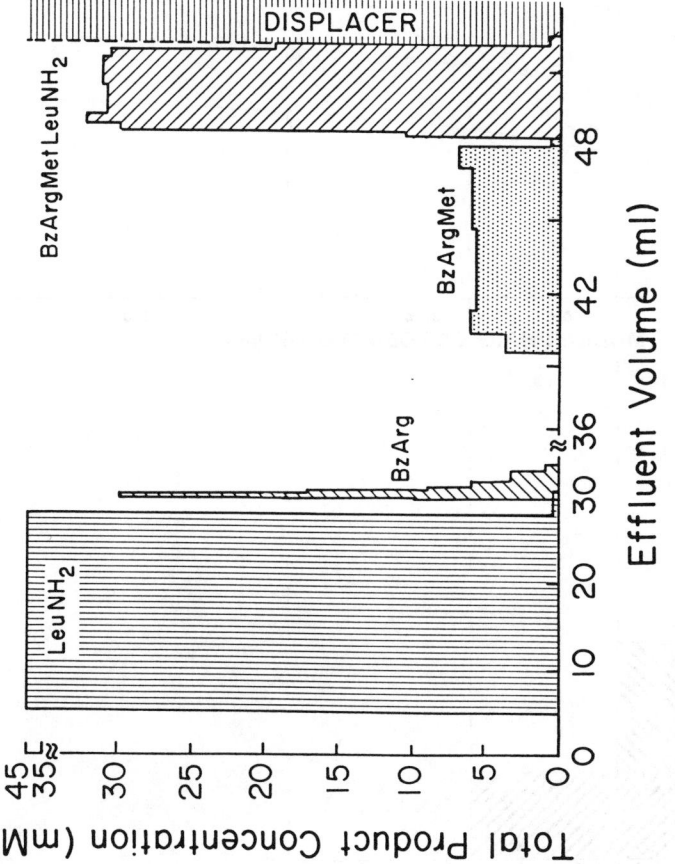

Figure 9. Displacement chromatogram obtained in the purification of Bz-Arg-Met-Leu-NH2 from 22 ml of feed. Column: 400 x 4.6 mm; C-18 silica; carrier: 0.1 M phosphoric acid, pH 2.2; displacer: 40 mM decyltrimethylammonium bromide in a solution of 15% (v/v) methanol in 0.1 M phosphoric acid, pH 2.2; temperature: 22°C and 50°C for feed and displacer respectively. (Reproduced with permission from reference #78, copyright, 1988, Gordon and Breach Science Publishers, Inc.).

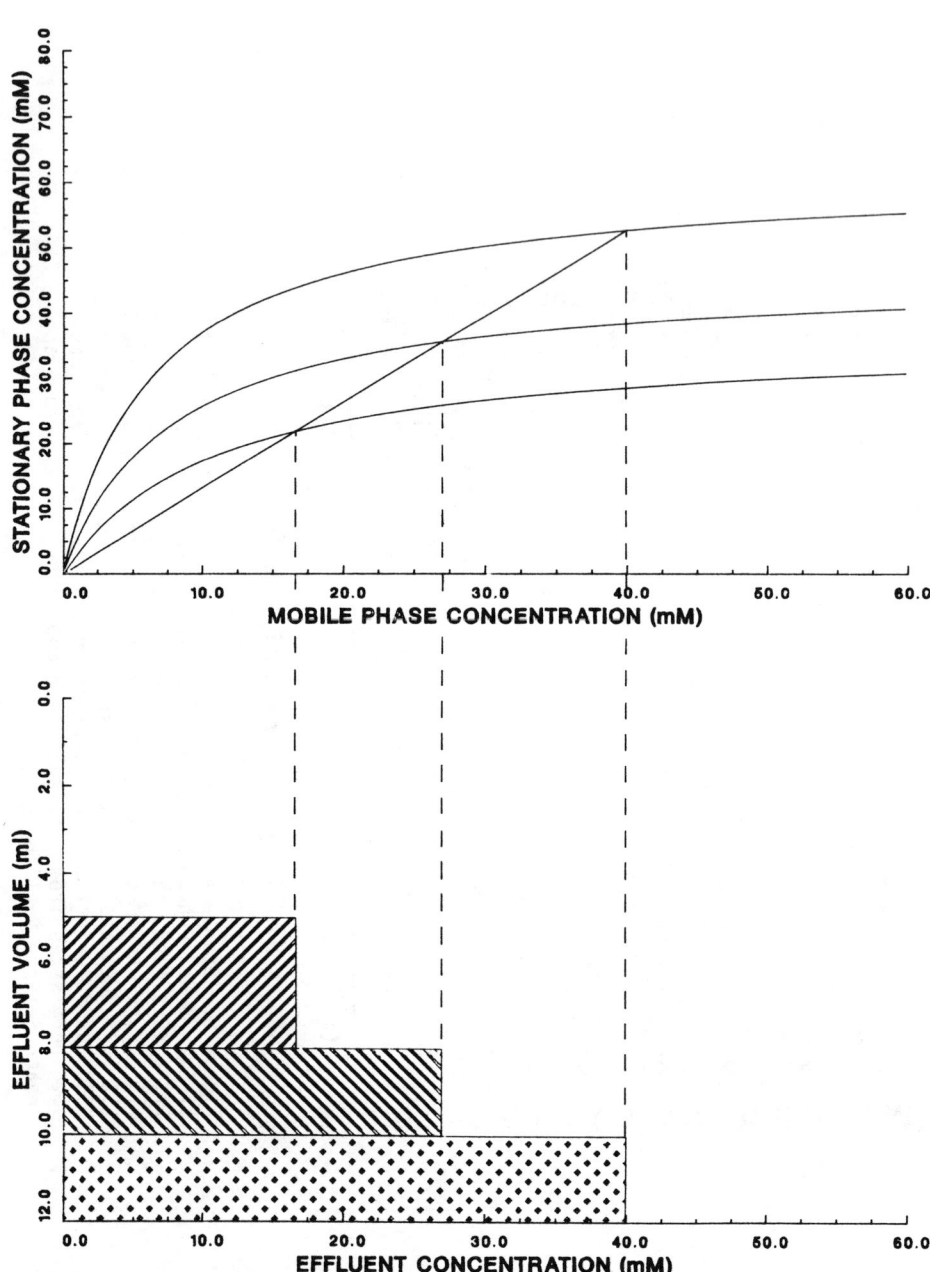

Figure 10.   Schematic representation of isotherms of the feed components and the displacer along with the operating line and the corresponding fully developed displacement train.

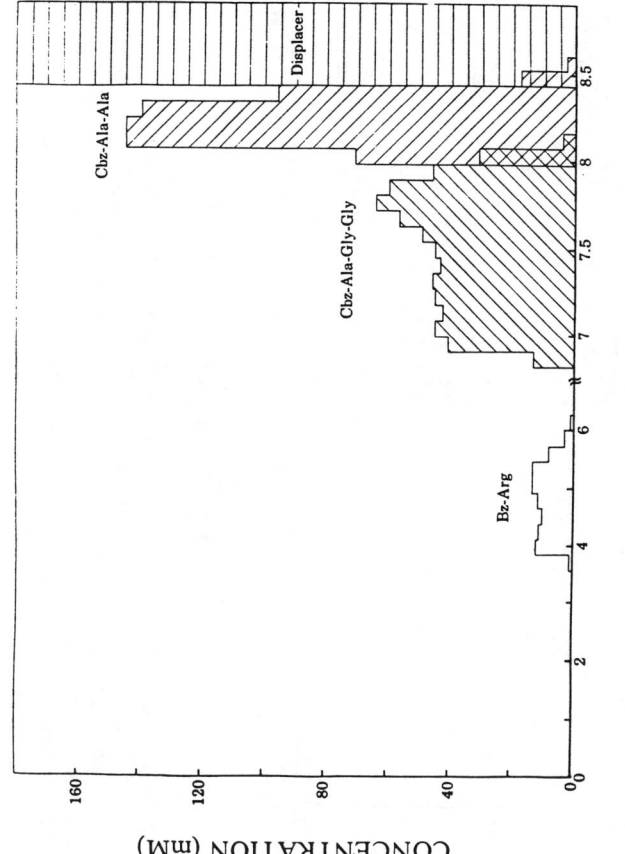

Figure 11. Displacement chromatogram of peptide mixture. Column: 250 x 4.6 mm Zorbax ODS; carrier: 40% (v/v) methanol in 50 mm phosphate buffer, pH 2.2; displacer: 30 mg/ml 2-(2-butoxyethoxy)ethanol BEE in carrier; flow rate: 0.06 ml/min; temperature: 45°C; feed: 7.2 mg Bz-Arg, 16.8 mg Cbz-Ala-Gly-Gly, and 18.4 mg Cbz-Ala-Ala. (Reproduced with permission from reference #82, copyright 1988, Elsevier Science Publishers B.V.).

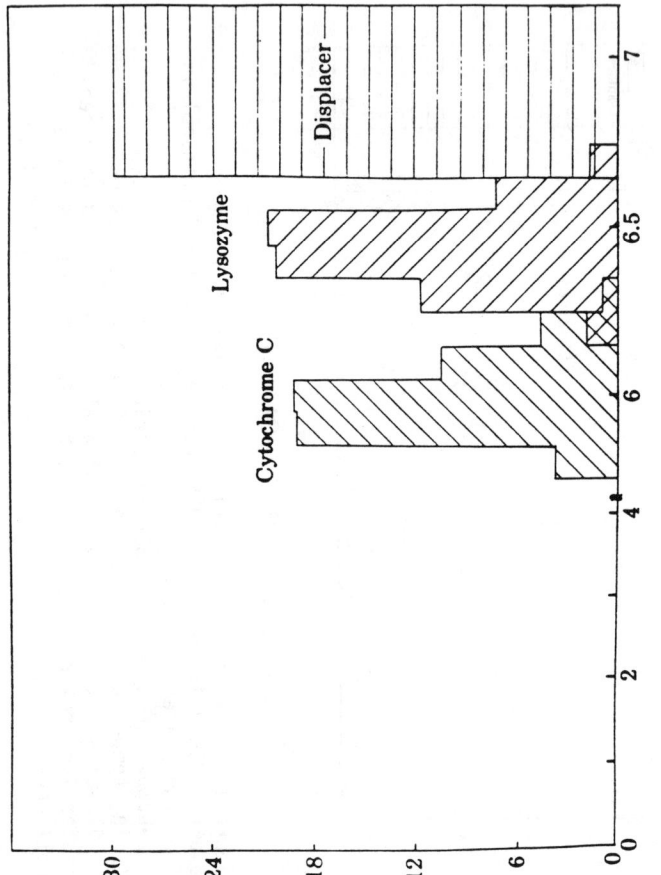

EFFLUENT VOLUME (ml)

CONCENTRATION (mg/ml)

Figure 12. Displacement chromatogram of protein mixture. Column: 250 x 4.6 mm WCX; carrier: 0.2 M sodium acetate, pH 6.0; displacer: 30 mg/ml Nalcolyte 7105 in carrier; feed, 6 mg each of cytochrome C and lysozyme; flow rate: 0.1 ml/min. (Reproduced with permission from reference #82, copyright 1988, Elsevier Science Publishers B.V.).

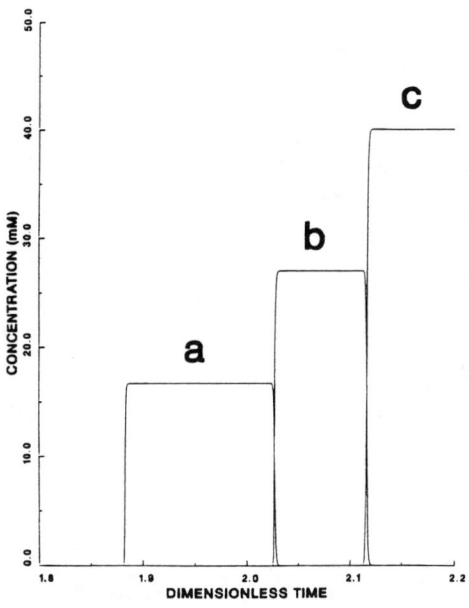

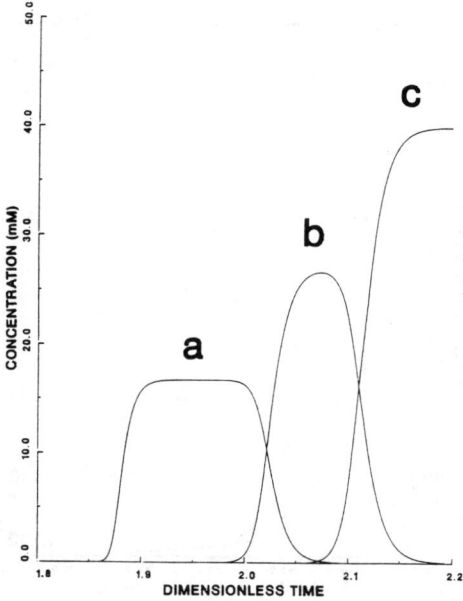

Figure 13. Effluent displacement profile obtained with a Stanton number of a) 1600 and b) 200 respectively. Simulation conditions are given in ref. 19. (Reproduced with permission from reference #19, copyright 1988, Elsevier Science Publishers B.V).

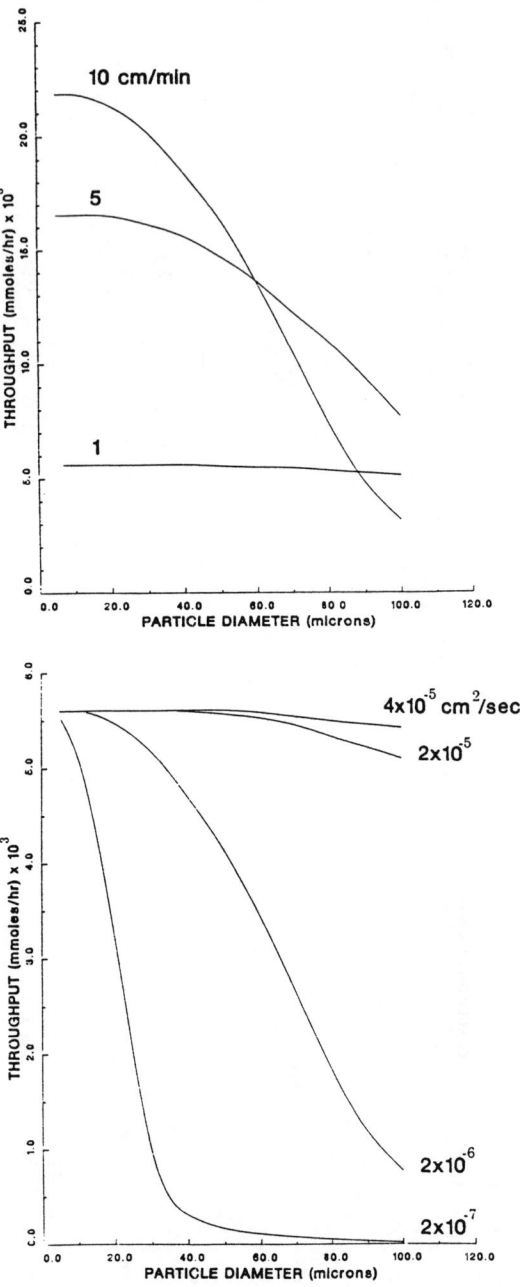

Figure 14.   The effect of particle diameter on product throughput at: a) a constant solute diffusivity of 2 x 10⁻⁵   cm2/sec, and b) a constant interstitial velocity of 1 cm/min.   (Reproduced with permission from reference #19, copyright 1988, Elsevier Science Publishers B.V.)

Chapter 12

Chromatographic Study of Aqueous Phase Adsorption on Activated Carbon Fiber with Bacterial Growth

Motoyuki Suzuki
Jin-Eon Sohn

When Activated Carbon Fiber (ACF) is applied to water treatment, high external surface area provides a good environment for attached growth of bacteria and then the cooperative effect of adsorption and bacterial reaction may be emphasized.

In order to clarify adsorption characteristics of organics on ACF in water, chromatographic measurement of homologues of organic acids, alcohols and glycols were made first. The adsorption equilibrium constants were determined from the first moments of the pulses and the second moments showed that the axial dispersion was a controlling machanism in broadening of the peaks.

Besides, a behavior of d-glucose pulses was examined in the ACF bed where bacteria growth was appreciable. Analytical solutions for chromatographic moments in the ACF bed with biological activities were obtained and comparison with the measurements showed that existence of bacterial activities greatly increases the retention of glucose pulses. The adsorption capacity of glucose in bacterial biomass was quantitatively determined.

Activated carbon fibers (ACF) are newly developed adsorbent and have several attractive features. ACF are produced not only from polymers such as viscose rayon, polyacrylonitrile and phenolic resin but also from meso-phase carbon derived from coal tar pitch. The latter precursor opens a wide possibility in commercial applications since the cost of ACF may become considerably reduced.

ACF's are supplied in many forms such as cloth, felt, paper etc. They are at present in a limited use in the field of gas treatments such as solvent recovery, dehumidification of air and removal of odor.

Diameter of the fibers ranges from 6 to 17 micrometers, which means that they take advantages of large specific external surface area and short internal diffusion distance. These advantages of ACF may be well utilized if they are used in liquid phase adsorption since the shorter diffusion distance in the fiber makes adsorption rate extremely rapid and thus even in liquid phase adsorption it is not necessary to worry about the intraparticle diffusion which usually becomes a rate controlling step of liquid phase adsorption.

From this point of view, trials were made to utilize ACF in removing bad odor and small quantities of carcinogenic organics found in drinking water sources such as trihalomethanes and trichloroethylene. The results of adsorption isotherms of halogenated organics have been presented (1,2).

Another advantage of ACF, a large external surface area, also provides an interesting possibility. ACF of fiber density 0.75 g/cc and 10 micrometers in diameter gives an external surface area of 0.54 m^2/g, which is more than hundred times as large as that of granular activated carbons. External surface of the activated carbons used in water treatment is often regarded as a good environment for habitation of bacteria and microbiological activities are considered to assist lengthening the dynamic adsorption capacity of adsorbent as has been known as biological regeneration effect(3). Then the large surface area of ACF may offer an ideal condition for utilizing cooperative function of adsorption and bacterial activity when it is properly used in water treatment.

The purpose of the present work is, by means of chromatographic technique, first to clarify fundamental adsorption characteristics of ACF in aqueous phase and secondly to see if the biochemical effect in the ACF system where bacterial activities coexist can be detected separately from adsorption effect of ACF.

Experimental

Materials. Activated Carbon Fiber (ACF) used throughout the experiment was the felt-type FT-15 provided by Kuraray Chemical Co. Fibers are 11 micrometers in diameter, 0.8 g/cm^3 in density, and 0.9 cm^3/g of micropore volume determined from nitrogen adsorption at the liquid nitrogen temperature. Pore size distribution showed that most of the micropores range between 8 to 10 Angstrom in radius as shown in Figure 1.

4.0 grams of ACF felt was cut into about 1.0 cm square patches and packed in a glass column of 1.2 cm in internal diameter and 18 cm in length. Felts were packed tightly to give a packing density of about 0.2 g/cm^3.

Chromatographic Measurement. Schematic diagram of the experimental apparatus is shown in Figure 2. Deionized water was introduced into the packed column. Careful evacuation pretreatment of the water was necessary to avoid air bubble generation in the fiber bed.

The column was kept in a water bath thermostated to 298K. Pulses of sodium chloride solution were used as a representative of nonadsorbable pulses, to determine void fraction in the bed. Alcohol homologues: methylalcohol, ethylalcohol, n-propylalcohol and iso-propylalcohol, carboxylic acid homologues: formic acid, acetic acid and propionic acid, and two glycols: ethyleneglycol and diethyleneglycol were used to see general characteristics of adsorption in the ACF bed when the chromatographic method is applied to aqueous phase adsorption. D-glucose pulse was introduced as an indicator of biodegradable organic materials.

By keeping the flow rate constant in the range of 0.2 to 2 cm^3/min, small amount of an aqueous solution of these materials was introduced as a pulse at the inlet of the column. The effluent of the column was sampled continuously by a fraction collector with one to five cubic centimeter aliquots and a total organic carbon analyser (TOC-10 B, Shimadzu Manufacturing Co.) or an electric conductivity meter (TOA CM-6A) were used to determine concentrations in the aliquots.

Growth of Bacteria in a Fiber Bed. Synthetic medium made from d-glucose (9 mg/l as TOC), urea (0.45 mg/l) and potassium dihydrogen-phosphate (0.1 mg/l) with tapwater was flown in the bed with the flow rate of 0.8 cm^3/min. Glucose concentration was kept low since dissolved oxygen in the inlet medium cannot become higher than the saturation limit of about 9 mg/l and it was undesirable if anaerobic condition was formed in the bed.

No special seeding of bacteria was made but spontaneous bacterial growth in the bed was confirmed from the decrease of exit concentrations of total organic carbon (TOC) and the increase of inorganic carbon (IC) as a result of bacterial respiration. Figure 3 shows the exit TOC and IC, which clearly indicates that appreciable bacterial growth started after half a day and after two days almost all of the glucose introduced in the bed was decomposed within the bed, suggesting bacterial accumulation had been continuing in this period. The columns after bacterial growth of one day (Point A in Figure 3), two days (B), three days (C) and seven days (E) were utilized for glucose pulse studies with inert water as a carrier stream.

The bacterial biomass grown in the bed was considered to be attached on the fibers since suspended bacteria could not stably grow in the bed because of the small residence time of medium solution. The amount of biomass was estimated by thermogravimetric analysis of the ACF sample. A small amount of ACF sample was taken out from a column, dried in room temperature and then set in the thermogravimetric balance where the sample was heated at a constant speed of 6K/min in an inert nitrogen stream. Weight decreases during heating up from 383 to 773K were considered to be the dry weight of bacteria existed on ACF. The results are given in Table I.

Table I. Amount of Bacterial Biomass on the ACF Bed
after Feeding of Culture Medium and the Moments
of Chromatographic Peaks of Glucose

Bed No.	Dry Weight of Biomass (g/g-ACF)	Volumetric Fraction of Wet Biomass, δ_{bio},* (cm3/cm3-bed)	$\ln(_0)$ (Z/u) (1/min)	mu_1 (Z/u) (-)
A	0.0070	0.141	0.11	27.5
B	0.0093	0.187	0.28	31.3
C	0.0156	0.312	0.14	24.8
D	0.230	0.460	0.32	32.5

* ACF packing density; 0.2 g/cm3-bed. Biomass of unit dry gram is assumed to occupy 100 cm^3 of wet volume.

Analysis of Chromatographic Moments

Moment   Analysis.   For   analyzing   chromatographic   pulses,   many
alternative methods have been proposed. Since basic equations involves
many   parameters   for   equilibrium   relations   and   rate   expressions,
analytical   solutions   in   the   time   domain   are   sometimes   too   much
complicated   and   even   in   the   case   when   analytical   solutions   are
available,   it is difficult to determine parameters   by   curve-fitting
the   analytical solutions with experimental data in the   time   domain.
Moments of the chromatographic elution curves, C_e, usually gives   only
two   global   parameters   that are the first   moment, μ_1,   defined   by
Equation 1 and the second central moment, μ'_2, defined by Equation   2,
but   experiments   under   different conditions   can   help   determining
parameters involved in the basic equations.

$$\mu_1 = \int_0^\infty C_e(t) \ t \ dt \ / \int_0^\infty C_e(t) \ dt \tag{1}$$

$$\mu'_2 = \int_0^\infty C_e(t) \ (t - u_1)^2 \ dt \ / \int_0^\infty C_e(t) \ dt \tag{2}$$

Moment   solutions have been obtained for several different systems   of
adsorption   chromatography   (4,5)   and   for   chemical   reaction
chromatography (6). One of the advantages of the moment method lies in
the   convenience of obtaining analytical solutions of moments   directly
from the Laplace-transformed basic equations (7).

Solutions   of   the   Moments.   Basic equations for   an   ACF   bed   with
biological   activities   on the surface of the fibers   are   derived   by
considering the following assumptions:
(1) Adsorption on ACF and on bacterial biomass are taken into account.
The   equilibrium relations of both systems are assumed as   Henry   type
for low concentration pulses.
ACF

$$q = K_a \ C \tag{3}$$

Biomass

$$q_{bio} = K_b \ C \tag{4}$$

where q_{bio} denotes the amount sorbed by biomass expressed in terms   of
unit dry weight of biomass.
(2)   For rate processes, intraparticle diffusion in the fibers can   be
considered to be rapid enough because of small diffusion distance   and
also   fluid-to-fiber   mass transfer may be neglected because   of   the
large   external surface area. Intrinsic adsorption rate on   adsorption
sites can be obviously neglected and then only longitudinal dispersion
has to be considered here as a dominant nonequilibrium parameter. Then
the   basic   equation   of   mass balance in the   fiber   bed   becomes   as
follows.

$$E_z \ d^2C/dz^2 - u_0 \ dC/dz - N_{ACF} - N_{bio} = (\varepsilon - \delta_{bio}) \ dC/dt \tag{5}$$

where N_{ACF} and N_{bio} are, respectively, mass transfer   rates   to   ACF
phase and to biomass phase. Void fraction in the bed without bacterial
growth, ε,   and   volumetric   fraction   of   bacterial   biomass, δ_{bio},
determine the actual void fraction in the bed, $(\varepsilon - \delta_{bio})$.
(3) Since adsorption on ACF is assumed as equilibrium adsorption, N_{ACF}

is given as,

$$N_{ACF} = (1-\varepsilon) \, \rho_b \, dq/dt = (1-\varepsilon) \, \rho_b \, K_a \, dC/dt \tag{6}$$

(4) In biomass phase, equilibrium adsorption may be assumed with the same reason as considered for ACF. Also, biological reaction is taken into account by a first order kinetics.

$$N_{bio} = \rho_{bio} \, \delta_{bio} \, dq_{bio}/dt + \rho_{bio} \, \delta_{bio} \, k_r C$$

$$= \rho_{bio} \, \delta_{bio} \, [\, K_b \, dC/dt + k_r \, C \,] \tag{7}$$

where ρ_{bio} denotes the density of the biomass. k_r represents the first order reaction rate constant on the biomass basis.
(5) In the above equations, fractional amount of biomass, δ_{bio}, changes due to the growth of bacteria but it is assumed that E_z or adsorption parameters are not affected by the change of δ_{bio}.
(6) Initial conditions and boundary conditions for impulse response are given as,

$$t = 0 : \quad C = 0 \quad \text{for } 0 < z < Z \tag{8}$$

$$z = 0 : \quad C = M \, \delta(t) \tag{9}$$

$$z = \infty : \quad C = 0 \tag{10}$$

The above set of equations are transformed into Laplace domain and the final transformed equation becomes,

$$E_z \, d^2C/dz^2 - u_0 \, dC/dz - A(p) = 0 \tag{11}$$

where

$$A(p) = (1-\varepsilon) \, \rho_b \, K_a \, pC + \rho_{bio} \, \delta_{bio} \, (K_b \, pC + k_r \, C)$$

$$+ (\varepsilon - \delta_{bio}) \, pC \tag{12}$$

with the boundary condition as

$$z = 0 : C = M \tag{13}$$

Then the solution is given as

$$C = M \, \exp(\, -\lambda \, (p) \, z \,) \tag{14}$$

where

$$\lambda \, (p) = u_0/2E_z \, (\sqrt{ \, 1 + 4 \, E_z \, A(p)/u_0^2 } - 1 \,) \tag{15}$$

The moments are related to the solution in Laplace domain. Zeroth relative moment, u_0, is given as

$$u_0 = C(0) \, / \, M = \exp \, (\, -\lambda \, (0) \, Z \,) \tag{16}$$

$$\lambda \, (0) = u/2E_z \, (/ \, 1 + 4E_z \, \rho_{bio} \, \delta_{bio} \, k_r \, / \, u^2 - 1 \,) \tag{17}$$

First absolute moment and second central moment are derived from

$$\mu_1 = -\lim_{p \to 0} C'(p)/C(0) \tag{18}$$

$$\mu'_2 = \lim_{p \to 0} C''(p)/C(0) - \mu_1^2 \tag{19}$$

These solutions are simplified in the case when $(4E_z/u_0^2)(\rho_{bio} \delta_{bio} k_r) \ll 1$. This condition is in most cases fulfilled since biochemical reaction rate constant is rather small. Then the approximate expressions for the moments are:

$$\mu_0 = \exp (- \rho_{bio} \delta_{bio} k_r Z/u) \tag{20}$$

$$\mu_1 = (Z/u)[(\varepsilon - \delta_{bio}) + (1-\varepsilon) \rho_p K_a + \rho_{bio} \delta_{bio} K_b] \tag{21}$$

$$\mu'_2 = (2Z/u)(E_z/u)[\mu_1/(Z/u)]^2 \tag{22}$$

These relations are further simplified for the ACF bed without bacterial activities by putting $\delta_{bio} = 0$. Then,

$$\mu_0 = 1 \tag{23}$$

$$\mu_1 = (Z/u)[\varepsilon + (1-\varepsilon) \rho_p K_a] \tag{24}$$

$$\mu'_2 = (2Z/u)(E_z/u)[\varepsilon + (1-\varepsilon) \rho_p K_a]^2 \tag{25}$$

Obviously, these equations can be easily derived from the general equations obtained for column chromatography (3,4).

Results and Discussions

First Moment of Non-reactive Pulses. First absolute moments of organics and sodium chloride on ACF are plotted in Figure 4. A linear relation is seen for each solute as is expected from Equation 23. First, from sodium chloride pulses, void fraction of the bed, ε, is determined to be 0.75. Then by applying Equation 23 to the relation between the first moment and Z/u, adsorption equilibrium constant, K_a, for each solute pulse is determined as shown in Table II.

Table II. Adsorption equilibrium constants of organics on ACF
(298 K. aqueous phase)

Organics	K_a (cm^3/g)	Organics	K_a cm^3/g)
Formic acid	24.9	Methyl alcohol	3.96
Acetic acid	94.3	Ethyl alcohol	20.1
Propionic acid	262	Propyl alcohol	118
Ethyleneglycol	7.18	Iso-propyl alcohol	89.4
Diethyleneglycol	10.4	D-glucose	66.8

Thus obtained K_a are plotted against molecular weight in Figure 5, which shows that K_a for alcohols and carboxylic acids are well correlated by a single line given below.

$$K_a = 0.175 \times 10^{MW/22.3} \tag{26}$$

K_a for glycols, however, lie far below this line. Also K_a of D-glucose on ACF was calculated as 66.8 cm^3/g. which is again far more smaller than the correlation when molecular weight of glucose, 180, is taken into account. This is because these organics are more hydrophilic compared with corresponding alcohols or acids (8).

D-Glucose Pulses on ACF with Bacterial Growth. Typical elution peaks of d-glucose and sodium chloride from the ACF bed without bacterial growth and the bed after culture medium was flown to grow bacteria on the fibers, are shown in Figure 6. Obviously sodium chloride pulses are little affected with the existence of bacteria in the bed but d-glucose peak in the bed with bacteria became smaller and came out later. From the zeroth moment (area) of the peaks of d-glucose for different flow rates in the Beds (A) to (E), reaction rate constants, k_r, were determined by plotting log μ_0 against Z/u. Then linear relations were established to each bed and the slopes determined correspond to ($\rho_{bio} \delta_{bio} k_r$), which are plotted in Figure 7 (triangles) against the volume fraction of bacterial biomass, δ_{bio}. δ_{bio} is obtained from the mass fraction of bacteria on the fiber determined from thermogravimetric analysis by multiplying the specific wet biomass volume (ρ_{bio}^{-1} = 100 cm^3/g dry biomass). From this plot the reaction rate constant, k_r, is obtained from an average slope by applying Equation 19. Thus determined reaction rate constant is 66 cm^3/g min.

From the first moments of d-glucose peaks, it is apparent that the existence of biomass increases the residence time of d-glucose in the bed. An additive reversible sorption effect on biomass is described by Equation 20. $\mu_1/(z/u)$ for each bed is plotted against d_{bio} in Figure 7 (solid circles). From the plot, apparent adsorption equilibrium constant on biomass, K_b, can be determined to be about 118 cm^3/g. This value may look rather large when compared with the adsorption equilibrium constant of d-glucose on ACF, K_a = 66.8 cm^3/g. But since biomass exists as a large wet volume, volumetric concentration factor in biomass is not at all large.

Second Moment of the Pulse. As is illustrated in Figure 6, most of the elution peaks were broadened and precise evaluation of second moments were rather difficult because of long tailings and low peaks. For the peaks with small K_a, this problem is minimized and then second moment analysis was possible though considerable error may be involved. In order to normalize second moments, H is defined as

$$H = \mu'_2/(2z/u) / [\mu_1/(z/u)]^2 \tag{27}$$

Then H becomes independent of K_a and K_b.

$$H = E_z/u^2 = (d_f/u)/Pe \tag{28}$$

Where d_f is the diameter of the fiber and Peclet number, Pe, is defined as $d_f u/E_z$.

Figure 8 shows H plotted against 1/u for sodium chloride, ethanol and ethyleneglycol. The slope of the regression line gives Pe =

0.0066. Also d-glucose pulses in the beds with and without bacterial activities gave the similar second moment result as included in Figure 8. Sakoda et al.(1) obtained Pe = 0.015 to 0.020 from a breakthrough experiment of trichloroethylene in an ACF bed whose void fraction was 0.83. The value determined here shows larger dispersion existed in the present case, which may be attributed to the maldistribution of flow in the bed because of tight packing of felt patches.

Conclusion

Chromatographic methods were applied to aqueous phase adsorption on the activated carbon fiber (ACF) bed. Adsorption equilibrium constants were determined from the first moments of the elution peaks for three carboxylic acids, four alcohols and two glycols of rather low molecular weights and d-glucose. Adsorption equilibrium constants for carboxylic acid homologues and alcohol homologues could be correlated with molecular weights by a single relation.

Behaviors of d-glucose pulses introduced in the ACF bed were an appreciable growth of bacteria existed by introducing culture medium for one to seven days, showed that glucose was reversibly sorbed by the bacterial biomass and the sorption equilibrium constant in the biomass was determined by assuming that sorption by the biomass occurs in parallel to adsorption on ACF.

From the second moments of the pulses, it was concluded that longitudinal dispersion is the main mechanism of peak broadening in the ACF bed. The longitudinal dispersion coefficient could be correlated by Peclet number based on the fiber diameter as Pe = 0.0066.

Legend of Symbols

$A(p)$: function defined by Equation 12, mmole/cm^3 min
C : concentration, mmole/cm^3
C_e : concentration at the exit, mmole/cm^3
d_f : diameter of the fiber, cm
E_z : longitudinal dispersion coefficient in the bed, cm^2/min
H : normalized second moment defined by Equation 27, -
K_a : adsorption equilibrium constant on ACF, cm^3/g
K_b : sorption equilibrium constant in biomass, cm^3/g biomass
k_r : first order reaction rate constant, 1/min
M : strength of pulse input, mmole min/cm^3
MW : molecular weight, g/mole
N_{ACF} : volumetric mass transfer rate from fluid to ACF, mmole/cm^3 min
N_{bio} : volumetric mass transfer rate from fluid to biomass, mmole/cm^3 min
p : Laplace parameter, 1/min
Pe : Peclet number defined as $d_f u/E_z$, -
q : amount adsorbed on ACF, mmole/g
q_{bio} : amount sorbed in biomass, mmole/g
t : time, min
u : superficial velocity of fluid in the bed, cm/min
Z : column length, cm
z : longitudinal position in the bed, cm

Greek Symbols

$\delta(t)$: Dirac's delta function, 1/min

δ_{bio}: volumetric fraction of biomass in the bed, -

ε : void fraction in the ACF bed, -

$\lambda(p)$: function defined by Equation 15, 1/cm

μ_0 : zeroth reduced moment, -

μ_1 : first absolute moment, min

μ_2 : second central moment, min^2

ρ_p : ACF fiber density, g/cm^3

ρ_{bio}: biomass density based on wet volume, dry g/cm^3 wet volume

Literature Cited

1. Sakoda, A.; Kawazoe, K.; Suzuki, M. 5th Asia Pacific Regional Water Supply Conference, SS8-1-1 (1985).
2. Sakoda, A.; Kawazoe, K.; Suzuki, M. Water Research, 21, 717 (1987).
3. Weber, W. J., Jr.; Friedman, L. D.; Bloom, R., Jr. Advances in Water Pollution Research, Proceedings 6th International Conference, p641 (1972).
4. Suzuki, M.; Smith, J. M. Advances in Chromatography, 13, 213 (1975).
5. Furusawa,T.; Suzuki, M.; Smith, J. M. Catalysis Review-Science and Engineering, 8, 43 (1976).
6. Suzuki, M.; Smith, J. M. Chem. Eng. Sci., 26, 221 (1971).
7. Suzuki, M.; J. Chem. Eng. Japan, 6, 540 (1973).
8. Belfort, G. et al.; AlChE J., 30, 197 (1984).

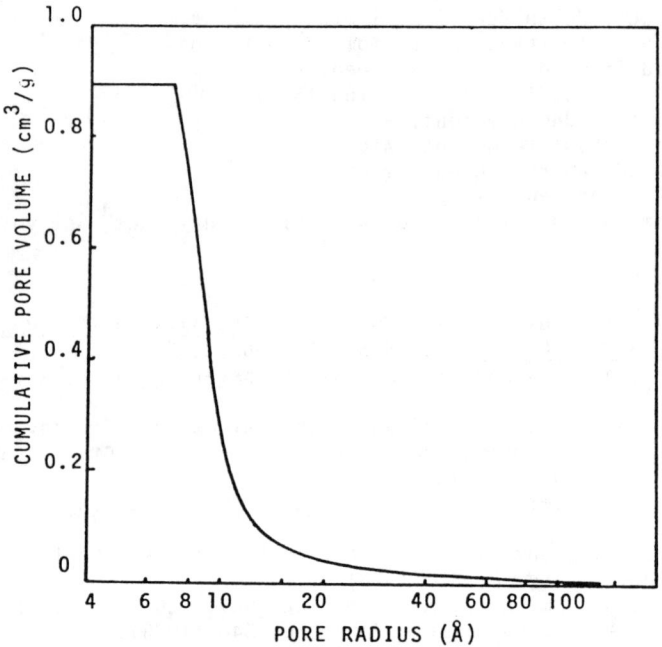

Figure 1. Pore size distribution of ACF determined from nitrogen adsorption at the liquid nitrogen temperature

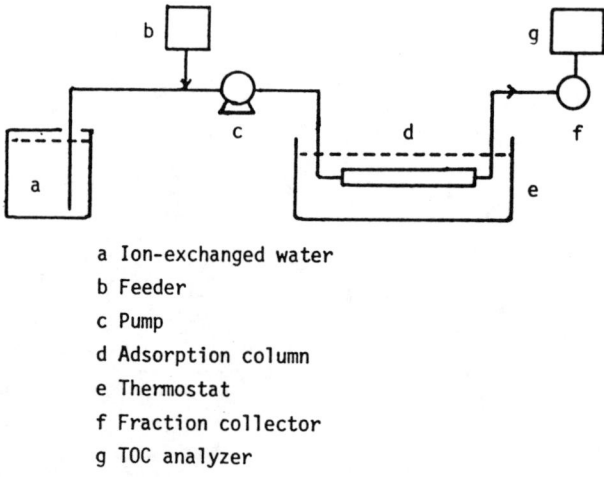

a Ion-exchanged water
b Feeder
c Pump
d Adsorption column
e Thermostat
f Fraction collector
g TOC analyzer

Figure 2. Schematic diagram of chromatographic experimental apparatus

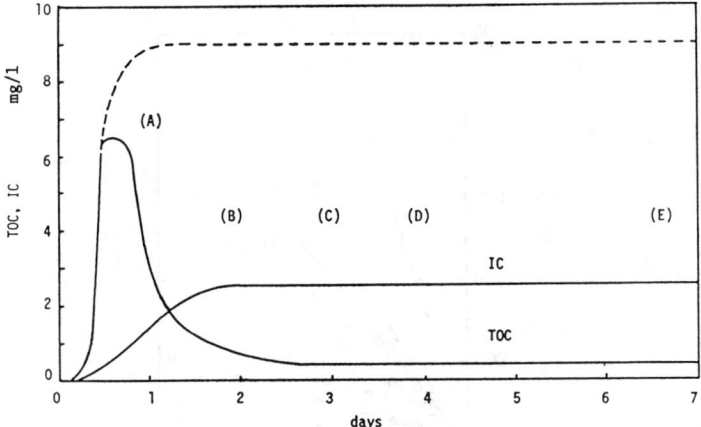

Figure 3. Change of TOC (total organic carbon) and IC (Inorganic carbon) in the effluent of the ACF bed with the culture medium feed flow.
Flow rate:0.8 cm^3/min, D-glucose concentration:9 mg/l as TOC

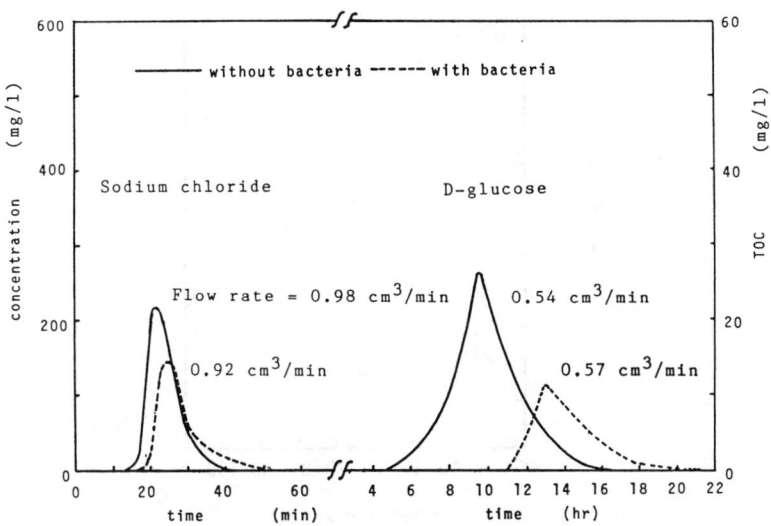

Figure 4. Typical chromatograms for the inpulses of sodium chloride and D-glucose for the ACF bed without bacteria (solid lines) and with bacteria (broken lines)

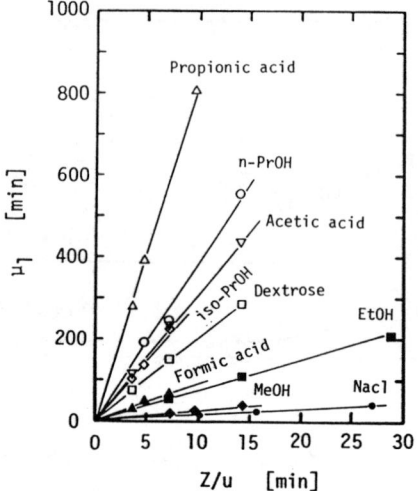

Figure 5. First moments of the chromatogams of the organics and sodium chloride for various flow rates

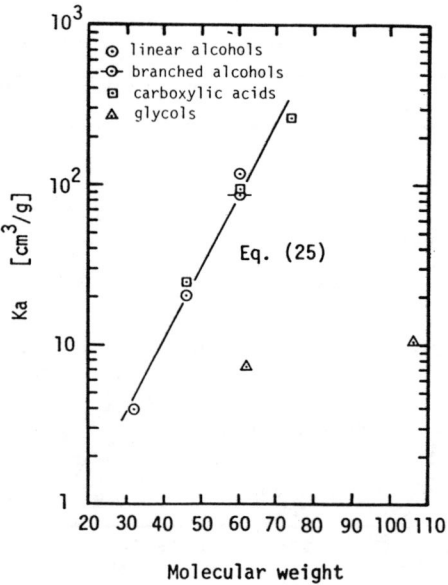

Figure 6. Adsorption equilibrium constants of the organics on activated carbon fiber, aqueous phase at 298 K. Solid line represents Eq.(25).

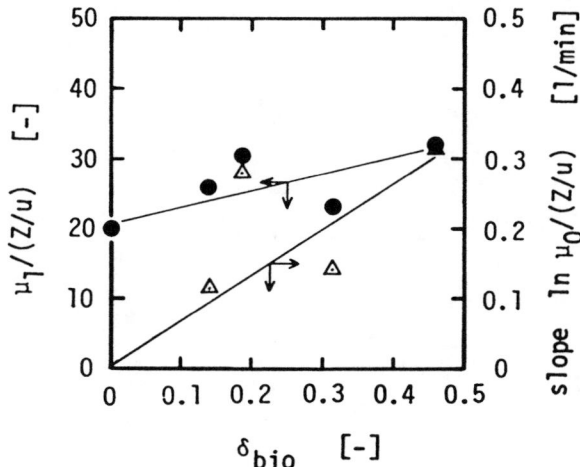

Figure 7. Slopes of the reduced zeroth moments (triangles) and slopes of the first moments (solid circles) in the ACF bed with bacterial growth against

the volumetric fraction of biomass in the bed

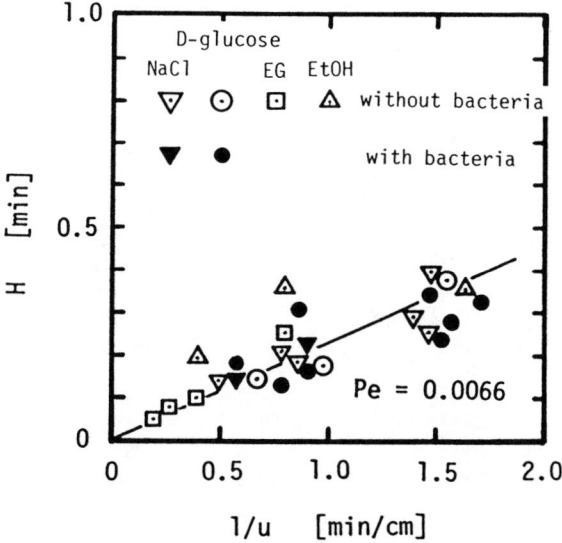

Figure 8. Dependence of the normalized second moments, H, on reciprocal velocity, 1/u

Chapter 13

Recovery of Bioactive Molecules and Recombinant Proteins from Heterogeneous Culture Media

Somesh C. Nigam
Henry Y. Wang

Recovery of biosynthetically derived products from the various culture broths is a complex engineering problem. Conventional broth processing involves a multistep scheme of non-specific unit operations which leads to significant loss of the desired bioproduct in the first few steps. We have been investigating the use of affinity adsorbents immobilized in reversible hydrogel beads or capsules for whole broth processing (1,2). The potential advantages of using immobilized affinity adsorbents include higher yield compared to the conventional bioseparation methods due to greater specificity. The reversible gel matrix may also render greater protection to the affinity ligand from fouling or enzymatic degradation. Earlier we reported the development of a mathematical model that describes antibiotic adsorption on immobilized affinity adsorbent beads (3). We are now interested in using similar immobilized affinity adsorbents for the recovery of recombinant proteins. In this study we examined whole broth extraction of a recombinant β-lactamase (M.W.-23,000) produced by <u>Bacillus stearothermophilus</u>. The applicability of the modeling approach developed earlier for small molecules (3), is now extended to describe macromolecule adsorption. Preliminary results indicate that this new approach to recover polypeptides and small protein products is feasible. However, bioproduct adsorption within these beads was found to be hindered by restricted diffusion and pore blockage effects. Elucidation of the complex interrelated mechanism of these effects and development of strategies to minimize their effect are required.

The development of recombinant DNA technology has provided the capability of producing large quantities of valuable bioproducts. Insulin, human growth hormone and interferons are some examples of important bioproducts which are being produced with this new technology. The development of a cost-effective bioseparation scheme to isolate these bioproducts in a highly purified form is one major hurdle for large scale commercialization of this technology. The desired bioproduct is generally present in very low concentrations(1-10 mg/ml in some cases) in extremely heterogeneous aqueous solutions. The bioproducts are also usually prone to chemical or enzymatic degradation. Conventional methods for downstream processing usually involve a multistep scheme consisting of fairly non-specific separation methods based on differences in molecular size, charge, hydrophobicity, etc. Such bioseparation schemes often lead to substantial loss of the desired bioproduct due to inherent non-specific nature of the involved unit operations. Other disadvantages of the conventional multistep bioseparation strategy are, additional capital/labor costs and long processing time which could further reduce process yield due to bioproduct degradation.

Affinity adsorption offers a promising alternative to the initial steps of a commercial biochemical separation scheme due to its high selectivity. It is a separation technique based on specific and reversible binding via biological interactions. Affinity adsorption is now being used in the form of chromatography for the final purification of bioproducts but is seldom used for processing more complex solutions such as whole culture broth. The severe decline in performance of affinity columns in the presence of suspended solids and fouling proteins has restricted its use to relatively clean semi-purified solutions. Furthermore, poor mechanical properties of the conventional affinity chromatographic packings have restricted this technique to small scale applications.

Recently we have suggested the use of immobilized affinity adsorbents to overcome some of these limitations (1,2). The immobilized affinity adsorbent consists of small, porous affinity adsorbent particles (such as adsorbent particles used for affinity chromatography) entrapped within a reversible hydrogel matrix (Figure 1). Reversible hydrogels can be dissolved by manipulating certain physico-chemical conditions such as ionic strength, temperature, etc. Calcium alginate, agarose and chitosan are used as model examples of the reversible hydrogels utilized for entrapment.(1) Most of the recombinant bioproducts being commercialized are either polypeptides or small proteins (M.W.<50,000). These bioproducts are expected to diffuse readily within the hydrogel since it consists of more than 90% water. (4) These beads can therefore be made relatively large (1-3mm) to ensure easy recovery from the whole broth at the end of the adsorption process. Large undesired macromolecules present in the whole broth are excluded from the hydrogel matrix because of the pore size restriction (molecules greater than M.W. 60,000 for calcium alginate matrix). Immobilized affinity adsorbent beads are able to combine the size exclusion of the gel matrix and specific affinity interactions to promote effective bioseparation. The exclusion of large molecular weight contaminants may also protect the affinity ligand and the adsorbed bioproduct from enzymatic degradation and fouling. The reversible hydrogel beads or capsules can be dissolved at the end of the adsorption process to recover the adsorbent particles with the bound bioproduct. Dissolution of the hydrogel matrix at the end of each adsorption cycle also ensures the removal of any fouling film which may have developed due to non-specific adsorption of large macromolecules to the hydrogel surface. The affinity adsorbents can be re-encapsulated with the reversible gel and be reused.

We have already reported the development of a mathematical model for studying the bioproduct adsorption via immobilized affinity adsorbent beads. (3) The model was verified for the adsorption of cycloheximide (a small molecular weight antibiotic) onto immobilized XAD-4 polymeric adsorbent. (2) The mathematical model was further utilized to carry out simulation studies to investigate the effect of various bead design parameters. For small molecular weight products, our experimental results have indicated higher adsorption capacity and greater protection from fouling contaminants using these immobilized affinity adsorbents. These affinity beads have also been effectively employed for on-line extraction of cycloheximide antibiotic during fermentation. (2,5)

We are now interested in using immobilized affinity adsorbents for the recovery of macromolecules obtained via recombinant DNA technology. The problems of separating labile protein bioproducts are compounded due to their high susceptibility to chemical and enzymatic degradation and the presence of numerous contaminating proteins in the broth. Many of these proteinaceous contaminants have size, charge, and hydrophobic characteristics similar to those of the desired bioproduct which limits the use of conventional, non-specific bioseparation methods. The use of a specific method based on affinity adsorption in the early stages of a bioseparation scheme has some definite advantages for recombinant protein isolation.

Model Experimental System.

In this work we have chosen the recombinant β-lactamase production by Bacillus stearothermophilus as a model experimental system. β-lactamase is a microbial enzyme (M.W.-23,000) which interacts specifically with a variety of β-lactam antibiotics and hydrolyzes them to biologically inactive products. Several researchers have already cloned and expressed β-lactamase in various host cells. (10) β-lactamase has been chosen as a model recombinant protein because of its relatively straight forward and sensitive assay procedure. (6) The recombinant strain of Bacillus stearothermophilus used in this study produces both extracellular and cell-bound β-lactamase activity. Extracellular β-lactamase is excreted into the culture medium at a concentration of about 3 I.U/ml (approx. 0.3-0.5 µg/ml). In the literature β-lactam antibiotics, such as cephalosporin C (7), ampicillin (8), and methicillin (9) have already been investigated as affinity ligands for small scale purification of β-lactamase. In this study we investigated the crude separation of extracellular recombinant β-lactamase from culture supernatant using immobilized affinity adsorbent beads. The mathematical modeling approach developed earlier was utilized to describe the simultaneous diffusion and binding effects in the beads.

Materials and Methods.

Organism: Various recombinant strains of β-lactamase producing Bacillus spp. were obtained from Professor Tadayuki Imanaka of Osaka University (Osaka, Japan). One strain of Bacillus stearothermophilus carries β-lactamase genes from both wild type and constitutive strains of Bacillus licheniformis 9945A on plasmid pTB90. This strain produces large quantities of both extracellular and cell bound β-lactamases. (10)

Chemicals. CNBr activated sepharose 4B, sodium alginate, 1-ethyl-3-dimethylaminopropyl carbodiimide, Tris base, cephalosporin C, 7-aminocephalosporanic acid and cephalosporinase (from Enterobacter cloacae) were obtained from Sigma Chemical Co. (St. Louis, MO.) Carboxyl controlled pore glass support was obtained from Electro-Nucleonics Inc. (Fairfield, NJ). CDI-glycophase support was purchased from Pierce Inc. (Rockford, IL.). Glutaraldehyde-activated silica gel was obtained from Serva Fine Biochemicals (Westbury, NY.).

Growth of Microorganisms. Bacillus stearothermophilus strain carrying the cloned β-lactamase genes were grown at 48°C in LG broth. The media used in this work is the same as described by Imanaka et al. (10) Kanamycin (5μg/ml) and tetracycline (3 μg/ml) were added to the culture medium to maintain the presence of the recombinant plasmid.

Production of Crude β-lactamase. The whole culture broth was centrifuged at 20,000 rpm for 20 minutes. The supernatant stored at 4°C with 0.01% NaN_3 as preservative constituted the crude β-lactamase preparation.

Enzyme Assay. β-lactamase was routinely assayed by a variation of iodometric assay described by Sargeant. (6) Since Ca^{++} ions were added to the crude β-lactamase preparation to ensure the integrity of the Ca-alginate gel beads, the use of a phosphate buffer for this assay was avoided because of its tendency to precipitate divalent cations. Instead, 0.025 mM HEPES, pH 7 buffer was used. One unit of β-lactamase activity is defined as the amount of enzyme that hydrolyzes 1mmol Penicillin G per minute at 25°C and pH 7.

Protein Assay. Protein was measured by the method of Bradford (11) using Bovine serum albumin as the standard.

Preparation of Affinity Adsorbents. Affinity adsorbent particles were prepared as follows: (1) CNBr-Activated Sepharose - Cephalosporin C: Cephalosporin C was coupled to CNBr activated sepharose as described previously. (7) The concentration of cephalosporin C in the coupling mixture was about 5 mg/ml. The coupled adsorbent was washed with 0.1M Tris HCl buffer (pH 8) for four hours to block the remaining active groups. The degree of substitution of the gel was estimated by the reaction of the adsorbent with iodine reagent (same as used in the β-lactamase assay) after incubation with excess cephalosporinase solution (4.8 units/ml in 0.1 ml phosphate buffer at pH 7; one cephalosporinase unit is defined as the quantity of enzyme required to degrade 1 mmol of cephaloridine per minute). (2) CDI-glycophase - Cephalosporin C: 3 gm of CDI-glycophase silica gel was added to 15 ml coupling solution containing about 120 mg.ml cephalosporin C in 0.1M, pH 9.8 carbonate buffer. The mixture was placed in a rotary evaporator to provide uniform mixing for seven hours at 4°C. The mixture was washed with 0.1M, pH 8, Tris HCl buffer for five hours to block the remaining active groups. (3) Carboxyl controlled-pore Glass - 7-aminocephalosporanic acid: Carboxyl CPG, a carboxyl terminal adsorbent matrix was coupled to 7-aminocephalosporanic acid through carbodiimide

condensation as described elsewhere. (12)   The
concentration of EDAC and 7-ACA in the coupling mixture
was 32 mg/ml and 12.5 mg/ml respectively

Preparation of Immobilized Affinity Adsorbent Beads.
1.5% sodium alginate solution was mixed with the
settled adsorbent particles. The immobilized affinity
beads were prepared by dropping this suspension into 2%
$CaCl_2$-$2H_2O$ solution in distilled water. The
immobilized affinity beads were cured for 5-6 hours and
were stored in 5mM HEPES buffer (pH 7) containing 22mM
$CaCl_2$. The membrane encapsulated adsorbents were made
according to the method previously described. (1)

**Characterization of Immobilized Affinity Adsorbent
Beads**. The settled volume of the adsorbent particles
was measured before making the immobilized affinity
adsorbent beads. The radius and the total volume of
these beads or capsules was determined by measuring the
volume of the buffer displaced by a given number of
beads in a volumetric flask. The adsorbent beads were
dabbed using a paper towel to remove any extra moisture
on the surface prior to the measurement. R_A represents
the amount of adsorbent contained in the immobilized
adsorbent beads. It was defined as the bulk adsorbent
volume divided by the total bead volume. The bulk
volume of the free adsorbent particles used to prepare
the immobilized affinity beads was measured by
centrifuging the adsorbent slurry in calibrated tubes.
R_A was found to be approximately 0.61. The overall
bead porosity was determined indirectly by measuring
the dilution of β-lactamase solution upon addition of
immobilized beads containing underivatized adsorbent
particles. Underivatized adsorbent particles had no
capability to adsorb β-lactamase. The overall bead
porosity was estimated to be 0.78 for our experiments.

Determination of the Adsorption Isotherm. The
adsorption isotherm of the free affinity adsorbent
particles was determined through batch experiments in
which a measured volume of settled adsorbent was
equilibrated with different concentrations of crude β-
lactamase preparation using an appropriate dilution
buffer. Adsorption isotherm can vary significantly
with small changes in environmental conditions. In
order to avoid uncertainty the isotherm was obtained by
diluting the crude β-lactamase preparation with the
same buffer which was employed in subsequent batch
experiments using immobilized adsorbent beads or
capsules.

Affinity Adsorption using Immobilized Affinity Beads.
102 beads were weighed and their combined volume was
measured using the buffer displacement method described
earlier. 6.5 ml of appropriately diluted β-lactamase

preparation was added to the beads in a 20 ml reaction vessel. The reaction vessel was mounted on a table top shaker to provide constant agitation. 50ml samples were periodically withdrawn from β-lactamase assay until equilibrium was reached. The temperature was kept constant throughout at 25°C.

Mathematical Modeling.

Figure 1 shows a schematic diagram of an immobilized affinity adsorbent bead. In our system the affinity binding reaction was found to be much faster compared to product diffusion in the bead. This justified the use of an equilibrium binding model considering only diffusional limitations. Graves and Wu (13) have developed a simple equilibrium model to describe bioproduct binding on affinity adsorbent particles in a well stirred system. Simplified equilibrium binding models such as Langmuir or Freundlich isotherms often fail to provide an adequate description due to the complex nature of these binding interaction. β-lactamase-cephalosporin system also displays a typical equilibrium adsorption isotherm which was found to be extremely sensitive to small differences in physico-chemical factors such as ionic strength. For the purpose of mathematical modeling, the experimentally derived adsorption isotherm was fitted by using a suitable mathematical function. The small adsorbent particles were assumed to be distributed uniformly within the hydrogel bead (Figure 1). The external mass transfer can be assumed to be negligible if the liquid is well stirred. (14) The diffusivity of the β-lactamase in the hydrogel matrix and within the adsorbent matrix was assumed to be close enough so that a single effective diffusivity could be defined to describe the diffusion in the composite matrix. In cases where the bioproduct diffusivity within the adsorbent matrix and the hydrogel entrapment matrix differs significantly the use of a bidisperse model may be necessary. (3) As an initial approximation the effective diffusivity was assumed to be constant throughout the adsorption process.

The product mass balance in a single immobilized affinity bead can be represented in the dimensionless form as:

$$\frac{1}{R^2}\frac{\partial}{\partial R}(R^2\frac{\partial C}{\partial R}) = \frac{\partial C_S}{\partial t} + \frac{\varepsilon\partial C}{\partial t} \tag{1}$$

The equilibrium relationship between the bound and the free product can be written as:

$$C_S = f(C) \tag{2}$$

In our analysis the experimental equilibrium data was curve fitted in parts by using a mathematical function of the following form:

$$C_S = R_A \alpha C / (\beta + C)$$

Initial and boundary conditions are:

$$\frac{\partial C}{\partial R} = 0 \quad ; \quad R = 0$$

$$C = C_b \quad ; \quad R = 1$$

$$C(R) = 0 \quad ; \quad t = 0$$

The bulk product concentration can be represented as follows:
In a batch reactor;

$$\frac{\partial C_b}{\partial t} = - \lambda \frac{\partial C}{\partial R}\bigg|_{R=1} \tag{3}$$

where
$$\lambda = \frac{4\pi N R_o^3}{V}$$

The parameters used to generate dimensionless variables are:

C_{bo}, initial batch concentration for concentration variables

R_o, bead radius for radial distance

R_o^2 / D_{eff}, for time

The mathematical model described above involves the solution of a non-linear partial differential equation. The physical parameters required to apply this model are; D_{eff}, R_o, ε, C_{bo}, V, N, and experimentally determined equilibrium isotherm obtained using free adsorbent particles. Effective diffusivity of the bioproduct within the bead matrix is a complex function of several physico-chemical factors such as ionic strength, pH, nature of contaminants, etc., and it needs to be experimentally determined. In our study effective diffusivity was used as an adjustable parameter to fit the batch experimental data. The model equations were solved numerically using a combination of implicit and explicit finite difference techniques. (14)

Results and Discussion.

Several affinity adsorbents were synthesized and tested for their ability to specifically bind β-lactamase from the crude preparation. Table I lists the ligand coupling characteristics and binding capacities of various affinity supports. CNBr-sepharose-ceph C and CDI-glycophase-ceph-C were found to be effective in separating β-lactamase from the crude preparation. Coupling of 7-aminocephalosporanic acid to carboxyl terminal controlled-pore glass through an aliphatic spacer arm resulted in a poor affinity adsorbent. This adsorbent demonstrated significant capacity to non-specifically bind contaminating proteins probably due to the hydrophobic effects amplified by the aliphatic spacer arm. We were unable to couple cephalosporin C in high enough ligand density using glutaraldehyde-activated silica to serve as a satisfactory affinity adsorbent in our initial sets of experiments.

Table I. Supports, Ligands and Binding Characteristics of Various Affinity Adsorbents Tested

AFFINITY MATRIX & SOURCE	AFFINITY LIGAND	COUPLING METHOD	BETA-LACTAMASE BOUND FROM 2.5 I.U/ml SOLUTION (I.U/ml ADSORBENT)
(1) CNBr-Sepharose (Pharmacia)	Ceph-C	CNBr	1.9
(2) CDI-Glycophase (Pierce)	Ceph-C	Carbonyldiimidazole	0.7
(3) Carboxyl-Controlled	7-ACA	Carbodiimide	~0.15
(4) Glutaraldehyde - Silica (Serva)	Ceph-C	Glutaraldehyde	?

Adsorption and Elution Conditions. The affinity adsorbents prepared were initially evaluated by performing conventional affinity chromatography using small quantities of the crude β-lactamase preparation. Previous studies described in the literature on the affinity chromatography of β-lactamase have utilized the following elution schemes; (1) Primary elution with pH 4-4.5, 0.1M acetate buffer and secondary elution with pH 7, 0.1M phosphate buffer (7,12), (2) primary elution with pH 7, 0.05M phosphate buffer and secondary elution with pH 7, 0.05M phosphate buffer containing 0.8 - 1M NaCl (9,12). The use of the first elution scheme was able to effectively bind β-lactamase but it also resulted in significant non-specific adsorption of contaminating proteins. It is suspected that the ion-exchange effects introduced because of the free carboxyl group on cephalosporin C molecule may have been responsible for the non-specific binding at low pH values. The use of elution method 2 significantly reduced the non-specific binding. The activity retained on the column could be eluted as a sharp peak by switching to the secondary eluent after eliminating most of the contaminating proteins. The elution scheme 2 was modified by replacing 0.05M

phosphate buffer with 5mM HEPES buffer in subsequent
experiments to ensure the integrity of the Ca-alginate
gel matrix which was used to prepare the immobilized
affinity adsorbent beads. More than 100 fold
purification was routinely achieved using this elution
method with both CNBr-sepharose-ceph C and CDI-
glycophase-ceph C adsorbents. All of the following
experiments were conducted using CNBr-sepharose-ceph C
since it showed the best characteristics among the
tested affinity adsorbents.

Adsorption Isotherm. The mathematical model used in
this study requires the determination of the
equilibrium isotherm for the adsorption of the
bioproduct using free affinity adsorbents. The results
of the batch equilibrium adsorption studies performed
with free CNBr-sepharose-cephalosporin C using two
different dilution buffers are shown in Figures 2a and
2b. The adsorption isotherms were found to be highly
non-linear in nature. The comparison of the adsorption
isotherms for active and cephalosporinase-degraded
adsorbent particles indicated significant non-specific
adsorption of β-lactamase on the degraded adsorbent
when a low ionic strength dilution buffer (5mM HEPES)
was used (Figure 2a). Increasing the ionic strength of
the crude β-lactamase preparation at different
dilutions by adding $CaCl_2$ (22mM) largely eliminated the
non-specific binding of β-lactamase on the sepharose
matrix. A large shift in the adsorption isotherm was
also observed under these conditions (Figure 2b). This
interesting observation demonstrates the significant
dependence of binding characteristics on small changes
in the physico-chemical conditions such as ionic
strength of the broth. Experimental determination of
the adsorption isotherm under actual operating
conditions is essential for the reliable design of an
affinity-based bioseparation scheme. In all of the
subsequent adsorption experiments involving immobilized
affinity adsorbents, we used 5mM HEPES with 22mM $CaCl_2$
at pH 7 as the diluting buffer.

Batch Studies using Immobilized Affinity Adsorbents.
The CNBr-sepharose-ceph C adsorbent particles were
entrapped in Ca-alginate hydrogel matrix to form the
immobilized affinity adsorbent beads. Dynamic batch
adsorption studies were conducted at three different
dilutions of the fermentation broth, namely 100%, 50%,
and 25%. The batch adsorption profiles were
numerically fitted using the mathematical model with
constant effective diffusivity as the only adjustable
parameter. Figure 3 shows the experimental and
numerically generated batch adsorption curves. The
mathematical model was able to adequately represent the
experimental data when the broth fraction was 25% or
50%. The model agreed well with the experimental data
only for the earlier time period of adsorption when the

undiluted crude enzyme preparation (100% broth fraction) was used. The diffusion process was prematurely halted after 225 minutes possibly due to pore blockage. The value of the diffusivity required to fit these adsorption curves was also found to increase as the crude preparation was diluted (1×10^{-6} cm^2/s, 6.95×10^{-7} cm^2/s and 3.47×10^{-7} cm^2/s for broth fractions 25%, 50%, and 100% respectively). These results indicate that a constant diffusivity model may not adequately represent the data when the solute effects become more dominant. Diffusivity itself depends on a number of factors such as molecular size, ionic strength, pH and the nature of the contaminants. The whole fermentation broth contains a mixture of small and large molecular weight contaminants. These contaminants could hinder the diffusion of the desired bioproduct in the bead. Large macromolecules could also adsorb onto the porous matrix to hinder and eventually block bioproduct diffusion. In our experiments with crude β-lactamase preparation both of these effects are evident from the concentration profiles. Yet another mechanism restricting bioproduct diffusion could involve the formation of an external fouling film on the bead surface. Significant restricted diffusion and pore blockage effects in hydrogel matrices have also been reported by several other researchers in more defined protein solutions. (15)

We have conducted scanning electron microscopic studies to examine the bead surface after premature cessation of bioproduct diffusion. The methodology for sample preparation has been described in detail by Chang. (16) The fresh - and the broth-exposed surfaces were compared as shown in Figure 4a and b respectively. The broth-exposed bead surface shows extensive fouling due to the formation of proteinanceous aggregates (Figure 4b). The external fouling film can introduce significant mass transfer resistance to bioproduct diffusion. However, at this stage it is difficult to estimate the extent of additional resistance provided by these aggregates. We suspect that at high broth fractions the bioproduct diffusion eventually stops due to the combination of micropore blockage within the matrix and the fouling of the external surface. The complex relationship between diffusivity, molecular size, pore blockage and nature of the fermentation broth will be the subject of our future studies. Such studies are needed to develop a more reliable mathematical description of binding and diffusion within the immobilized affinity beads and investigate various strategies to curb the fouling effects. The efficient reuse of the affinity adsorbents will be crucial for the economic use of the proposed bioseparation scheme. The pore blockage and fouling observed in our experiments also indicates the utility of having the reversible hydrogel matrix for protecting the affinity ligand from these effects. The reversible gel, which guards against fouling contaminants, can be dissolved after each adsorption cycle thus allowing recovery of the adsorbents and regeneration of the immobilized affinity beads.

Conclusions.

The use of immobilized affinity adsorbent beads presents a novel approach for whole broth processing. Bioseparation systems based on immobilized affinity beads have the potential to enhance the overall extraction yield by utilizing specific affinity interactions during the initial steps of bioseparation. We have used the immobilized affinity adsorbent beads to extract recombinant β-lactamase from the fermentation broth of <u>Bacillus stearothermophilus</u>. The experimental results with the chosen model system indicate the practicality of this approach for the recovery of recombinant bioproducts. A mathematical model has been utilized to investigate the simultaneous diffusion and binding reaction within these beads or capsules. The value of effective β-lactamase diffusivity required in the mathematical model to fit the experimental data was found to increase with broth dilution. The constant effective diffusivity assumption applied only for very diluted broths. This indicates that various contaminants present in the whole broth could severely hinder bioproduct diffusion through the gel matrix. The use of undiluted broth lead to premature cessation of the bioproduct diffusion possibly due to pore blockage. Greater understanding of the restricted diffusion and pore blockage effects is required to quantify these phenomena. Examination of the bead surface using scanning electron microscopy revealed the formation of an irregular fouling film consisting of proteinaceous aggregates. The exact mechanism of pore blockage is however not clear at this stage.

Acknowledgement

We acknowledge the financial support from the National Science Foundation (ECE-86-12969 and CBT-88-1892).

Notation

R   Radial distance, dimensionless
C   Bioproduct concentration within the bead, dimensionless
C_b Bioproduct concentration in the batch, dimensionless
C_s Adsorbed bioproduct concentration within the bead, dimensionless
t   Time, dimensionless
R_A Bulk volume of entrapped adsorbent particles/bead volume
α, β Constant parameters used to fit the equilibrium curve
ϵ Average bead porosity
V   Batch volume, ml
N   Number of beads per batch
C_{bo} Initial batch concentration, I.U/ml
R_o Bead Radius, cm
D_{eff} Effective Diffusivity, cm^2/s

References

1. Nigam, S.C., Sakoda, A., and Wang, H.Y.,
    Biotechnology Progress 4, 166 (1988).
2. Wang, H.Y. and Sobnosky, K., "Design of a New
    Affinity Adsorbent for Biochemical Product
    Recovery" in ACS Symposium Series, No. 271, ed. D.
    LeRoith, J. Shiloach, and T.J. Leahy,   (1985).
3. Nigam, S.C. and Wang, H.Y., "Mathematical Modeling
    of Bioproduct Adsorption using Immobilized Affinity
    Adsorbents," in ACS Symposium Series 314, Ed. J.A.
    Asenjo and J. Hong, (1986)
4. Tanaka, H., Matsumura, M., and Veliky, I.A.,
    Biotech. Bioeng., 26,53, (1984).
5. Wang, H.Y., Annals of the New York Academy of
    Sciences, 413, 313 (1983).
6. Sargeant, M.G. Bacteriol, J. 95, 1493 (1968)
7. Crane, L.J. Bettinger,G. E. and Campen, J.O.
    Biochem, Biophys. Res. Commun., 50, 220 (1973)

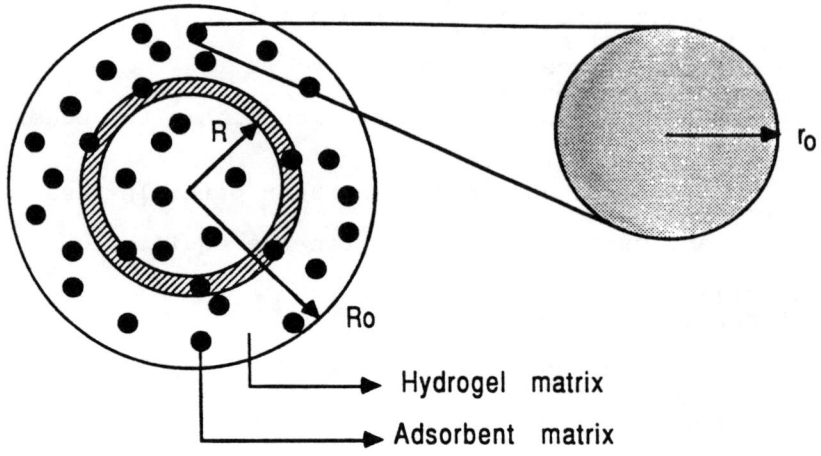

Figure 1.   A schematic diagram of the immobilized adsorbent
concept.

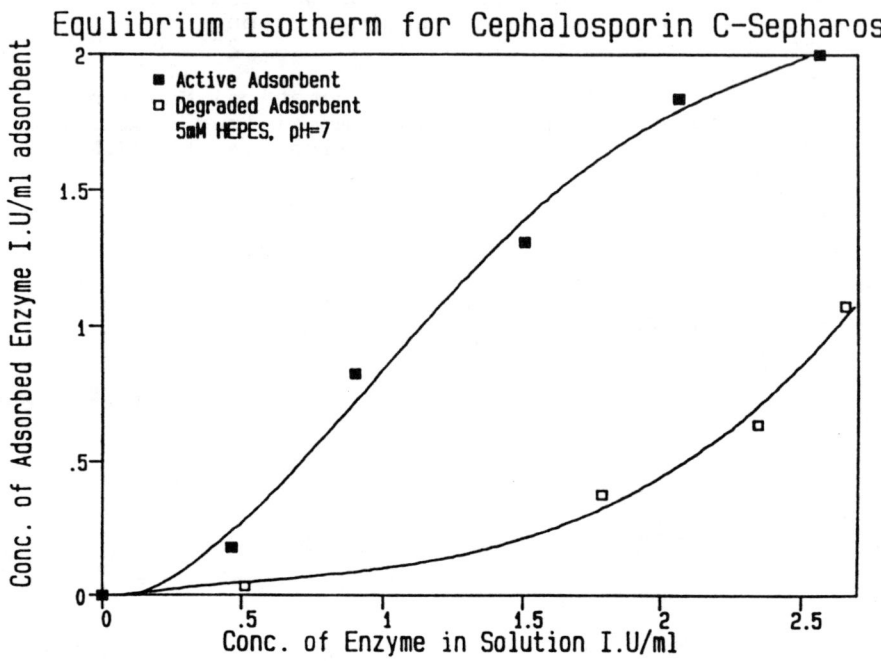

Figure 2a.   Adsorption isotherm for cephalosporin C- sepharose
using HEPES buffer (5 mM) - low ionic strength.

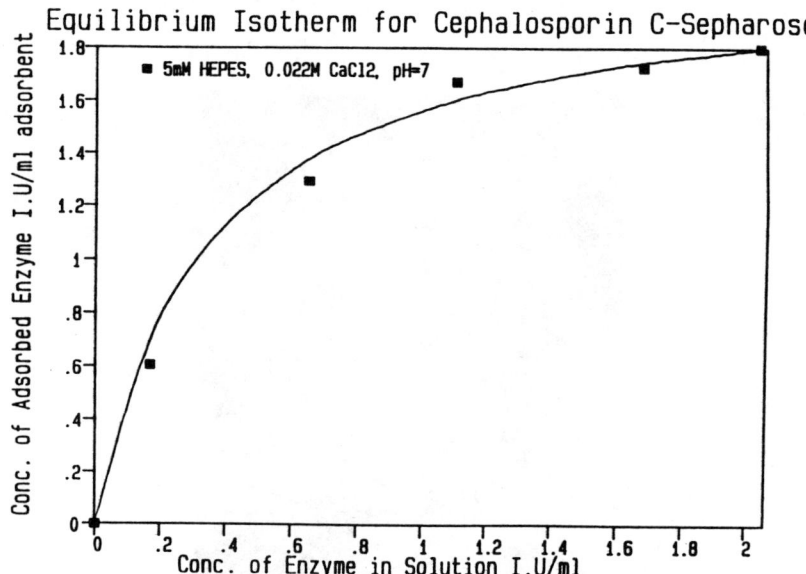

Figure 2b. Adsorption isotherm for cephalosporim C- sepharose using HEPES buffer (5 mM) plus 22 mM of CaCl$_2$ high ionic strength.

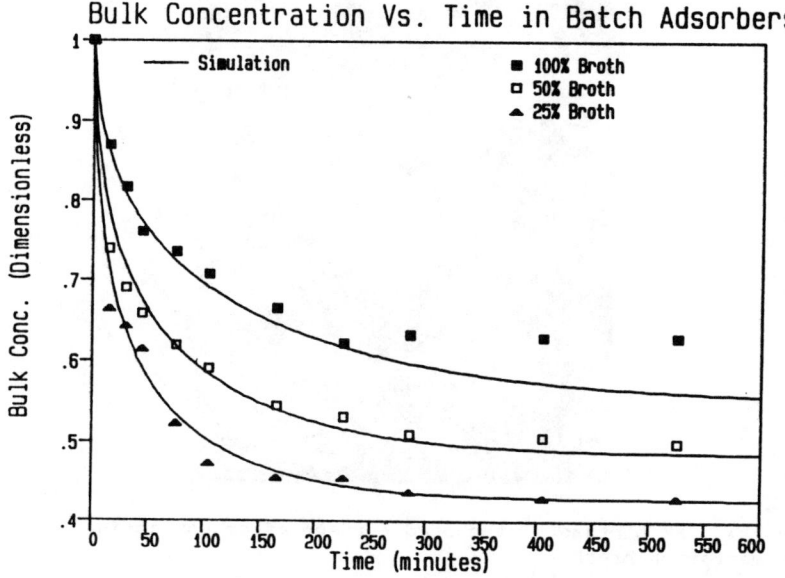

Figure 3. Computer simulated results in comparison with the experimental batch adsorption data.

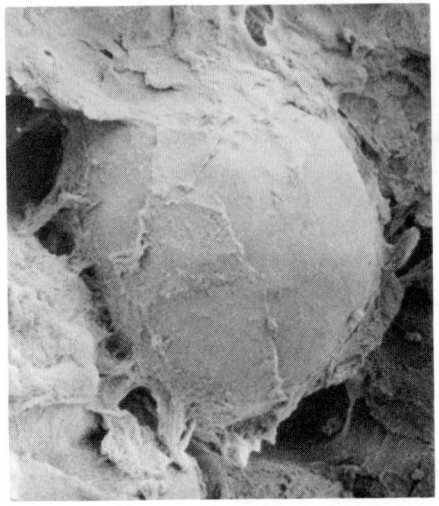

Figure 4a.   SEM picture of the freshly prepared bead surface.

Figure 4b.   SEM picture of the bead surface after contacting the fermentation broth.

INDEX